Brachiopods from the Maastrichtian–Danian boundary sequence at Nye Kløv, Jylland, Denmark

MARIANNE BAGGE JOHANSEN

Johansen, M. B. 1987 10 31: Brachiopods from the Maastrichtian–Danian boundary sequence at Nye Kløv, Jylland, Denmark. *Fossils and Strata*, No. 20, pp. 1–99, Oslo. ISSN 0300-9491. ISBN 82-00-02558-6.

The articulate brachiopod fauna from the Maastrichtian–Danian boundary sequence at Nye Kløv, Northern Jylland, Denmark has been studied to test terminal Cretaceous mass extinction models in a major marine benthic invertebrate group. Taxonomy, stratigraphy, density, diversity and palaeoecology of the brachiopods are examined. The results show an almost complete extinction of the brachiopod fauna at the boundary which is compatible with that reported for pelagic nannofossils and foraminifers. The extinction is abrupt, and there is no warning in the form of decreasing diversity or early extinction of specialized groups. The basal Danian clay bed, the Fish Clay, the base of which marks the Maastrichtian–Danian boundary, contains thirteen Maastrichtian brachiopod species. They are probably reworked from the uppermost Maastrichtian. The basal few metres of the Danian are devoid of brachiopods, probably reflecting soupy unstable substrate conditions and absence of suitable hard substrate for the brachiopods. A Danian brachiopod fauna starts almost as abruptly as the Maastrichtian fauna disappeared. The new fauna is similar to the Maastrichtian in terms of density and diversity. The Danian brachiopod fauna is dominated by pedicle-attached forms with wide substrate tolerances. The highly specialized reclining, secondarily free-living species characteristic of the Maastrichtian fauna are not represented in the Lower Danian. Six species, one inarticulate and five articulate, are common to both faunas, and all of these are long-ranging, morphologically non-specialized forms. There are 27 species, representing 16 genera and 9 families in the Upper Maastrichtian, and 29 species, representing 13 genera and 8 families in the Lower Danian. In the Danian at least 23 species and 2 genera are new compared to the Upper Maastrichtian. At least 1 genus and 5 species in the Lower Danian are hitherto undescribed. They are: *Cryptopora perula* n.sp, *Terebratulina kloevensis* n.sp., *Rugia flabella* n.sp., *Rugia latronis* n.sp., and *Gwyniella persica* n.gen. et n.sp. □ *Brachiopoda, taxonomy, chalk, Maastrichtian–Danian boundary, Nye Kløv, Denmark.*

Marianne Bagge Johansen, Geological Museum, University of Copenhagen, Øster Voldgade 5–7, DK-1350 Copenhagen K, Denmark; 1986 04 10, revised 1987 04 06.

Contents

Introduction 3
  Acknowledgements 3
The Maastrichtian–Danian boundary in Denmark 3
The Nye Kløv locality 5
  Stratigraphy 6
Material and methods 7
  Methods 7
  Material 9
  Preservation 9
  Brachiopod terminology, measurements
  and morphology 9
Systematic descriptions 10
Rhynchonellidae 10
  Genus indet. 10

Cretirhynchia sp. 10
Cryptoporidae 11
  Cryptopora perula n.sp. 11
Terebratulidae 12
  Neoliothyrina? sp. 12
  Carneithyris subcardinalis 13
  Carneithyris sp. 13
Cancellothyrididae 14
  Terebratulina chrysalis 14
  Terebratulina faujsaii 15
  Terebratulina gracilis 16
  Terebratulina longicollis 16
  Terebratulina cf. *longicollis* 18
  Terebratulina kloevensis n.sp. 18

Terebratulina aff. *rigida* 20
Gisilina jasmundi 21
Rugia acutirostris 22
Rugia tenuicostata 22
Rugia flabella n.sp. 23
Rugia latronis n.sp. 24
Rugia sp. 25
Meonia semiglobularis 26
?Megathyrididae 26
Gwyniella n.gen. 26
Gwyniella persica n.sp. 26
Argyrotheca bronnii 27
Argyrotheca aff. *bronnii* 28
Argyrotheca coniuncta 28
Argyrotheca danica 29
Argyrotheca stevensis 30
Argyrotheca aff. *stevensis* 30
Argyrotheca hirundo 33
Argyrotheca dorsata 34
Argyrotheca vonkoeneni 36
Argyrotheca cf. *faxensis* 38
Argyrotheca armbrusti 38
Platidiidae 39

Platidia sp. 39
Aemula inusitata 39
Scumulus inopinatus 40
Scumulus? sp. 41
Dallinidae 42
Kingena pentangulata 42
Dalligas nobilis 42
Dalligas sp. 43
Terebratellidae 43
Magas chitoniformis 43
Family uncertain 44
Leptothyrellopsis sp. 44
Changes in the brachiopod faunas at Nye Kløv 45
The washing residue curve 45
The fauna in the washing residue 45
The benthos as a substratum for the brachiopods . . 46
The brachiopod fauna 46
Brachiopod extinctions across the Maastrichtian–
Danian boundary 50
Summary 53
References 53
Plates 60

Introduction

The brachiopods from the lowermost Paleocene Danian Stage have hitherto been treated only cursorily. The more important contributions on Danish material are by Koenen (1885), Posselt (1894) and Nielsen (1911, 1914, 1920, 1921, 1928 and 1937) and, of comparatively recent date, Asgaard (1968, 1970) and Johansen (1982).

The uppermost Cretaceous (Maastrichtian) chalk brachiopods are, in contrast, well known and have been described from many localities in northwestern Europe. Among early workers are Davidson (1854), who studied the British Upper Cretaceous brachiopods, Bosquet (1859), who studied the brachiopods from the Limburg area of the Netherlands, Schloenbach (1866), who studied material from Northwest Germany, and de Morgan (1883), who studied material from the Normandy. Of more recent date are the papers by Steinich (1963, 1965, 1967, 1968a, 1968b) who studied the brachiopods from Northwest Germany, Popiel-Barczyk (1968, 1973, 1977) Panow (1969) and Bitner & Pisera (1979) who studied material from the Polish Upper Cretaceous, and Surlyk (1969, 1970a, 1970b, 1971, 1972, 1973, 1974, 1979a, 1982, 1983) who described the brachiopods from the Upper Campanian–Maastrichtian of Northwest Germany and Denmark. In addition, Asgaard (1975) and Owen (1970, 1977) have recently described the Upper Cretaceous brachiopods of Norfolk, England (a manuscript by Johansen & Surlyk is also in preparation), and Johansen (1986) has studied the evolution of the Middle Coniacian – Lower Maastrichtian brachiopods of Lägerdorf and Kronsmoor, Northwest Germany.

The micromorphic brachiopods are of great biostratigraphic importance, and a biostratigraphic zonation of the Upper Campanian–Maastrichtian of Northwest Europe has been established by Steinich (1965) and Surlyk (e.g. 1970b, 1984).

The chief objectives of the present paper are to describe the brachiopod fauna from the chalk of the Nye Kløv section, northern Jylland, Denmark, with the main emphasis on the forms from the Lower Danian. In the Lower Danian of Nye Kløv 35 species are found, six of which are restricted to the basal Danian clay bed, the Fish Clay. These six are common in the immediately underlying Upper Maastrichtian chalk, are not present in the succeeding Lower Danian strata, and have most probably been reworked from the Upper Maastrichtian. Six species are common to the Maastrichtian and the Danian and represent forms that have crossed the boundary. The remaining 23 species appear for the first time in the Danian, and of these, five species are here described as new.

Another objective of the present paper is to analyse the extinction patterns of the brachiopod fauna at the Maastrichtian–Danian boundary. The Nye Kløv locality has recently been proposed as a substitute for Stevns Klint as stratotype for the Cretaceous–Tertiary transition due to possibly more continuous sedimentation across the boundary at Nye Kløv (Surlyk 1983).

Acknowledgements. – This work is part of my Ph.D. thesis at the Institute of Historical Geology and Palaeontology, University of Copenhagen, supervised by Professor Finn Surlyk, Geological Survey of Greenland, Copenhagen, to whom I want to express my sincere thanks for help and discussions during the course of the work. For practical assistance I thank Dr. Erik Thomsen, Institute of Paleoecology, Århus, Denmark, Mr. Ole Bang Berthelsen, Mr. Henrik Egelund, Mr. Jens Fuglsang Nielsen, Mr. Jan Ågaard and Dr. Eckart Håkansson, all of the Geological Institute, Copenhagen, and Mrs. Bodil Sikker Hansen and Mrs. Vibber Hermansen, both of the Geological Survey of Greenland, Copenhagen. For assistance during the final stages of the manuscript I want to thank Mr. Christian Rasmussen, Mrs. Ragna Larsen and Mrs. Annemarie Brantsen, all of the Geological Museum, University of Copenhagen.

Dr. Walter Kegel Christensen, Geological Museum, University of Copenhagen, Professor Finn Surlyk, Geological Survey of Greenland, Copenhagen, Dr. John Hurst, B.P. Petroleum Ltd., London, and cand. scient. John Miller, Institute of Historical Geology and Palaeontology, Copenhagen, have critically read the manuscript and offered many helpful suggestions.

I also want to thank Professor Albert J. Rowell, Department of Geology, University of Kansas, Lawrence, USA, for refereeing the manuscript, and Dr. Stefan Bengtson, Institute of Palaeontology, University of Uppsala, Sweden, for many helpful comments.

Finally I want to express my heartfelt thanks to my husband and daughters for their questions and fingerprints.

During the course of the present work, the author has been employed by the University of Copenhagen. The Danish Natural Science Research Council has paid for the printing of this paper.

The Maastrichtian–Danian boundary in Denmark

The boundary between the Mesozoic and Cenozoic Eras is placed at the Maastrichtian–Danian boundary. A number of outcrops exposing the boundary are present in Denmark. The most important of these localities are Bjerre, Danian, Eerslev, Kjølbygård, Nye Kløv and Vokslev in Jylland, in addition to Karlstrup and Stevns Klint (Kulsti Rende, Holtug, Højerup and Rødvig) in Sjælland (Fig. 1).

The Cretaceous–Tertiary boundary is traditionally biostratigraphically defined (e.g., Forchhammer 1825; Ravn 1903; Rosenkrantz 1924, 1937, 1940; Birkelund & Bromley 1979). In the majority of localities the boundary is marked also by a global hiatus which, in most places, includes both the Upper Maastrichtian and the Danian. These stratigra-

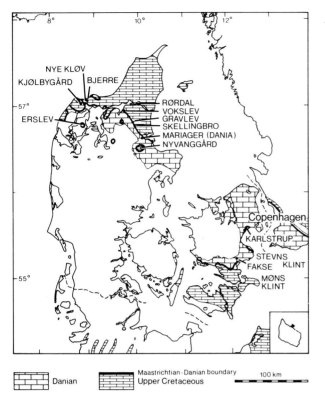

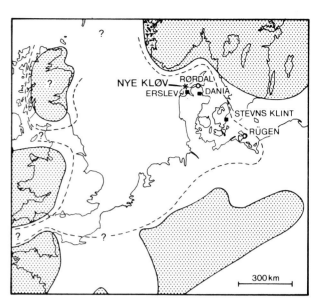

Fig. 2. Palaeogeography of the Maastrichtian sea in NW-Europe. The dashed line indicates the approximate extension of the chalk facies in the Middle Maastrichtian. The dotted areas are stable land masses. (After Håkansson, Perch-Nielsen & Bromley 1974).

Danian		Maastrichtian-Danian boundary
		Upper Cretaceous

100 km

Fig. 1. Geological map of Denmark showing distribution of Upper Cretaceous, Danian and Maastrichtian–Danian boundary strata, at the base of the Quaternary.

phic gaps reduce the possibility of studying the precise nature of extinction patterns across the boundary. Good boundary sections are characterised by a number of features, including more or less continuous sedimentation and few or no facies changes across the boundary, accompanied by minimum diagenetic dissolution and transport destruction of skeletal material. In addition, high density and diversity of studied fossil groups are important (Surlyk & Johansen 1984).

Due to heterogeneous lithological development of the Maastrichtian and Danian boundary strata at different localities in Denmark (Floris *et al.* 1971), lithological correlations have not been satisfactory, even though characteristic flint horizons, marly layers (e.g. the Kjølbygård Marl from the Upper Maastrichtian) and the Fish Clay (Christensen *et al.* 1973) and its equivalents, located at the boundary, have been shown to be useful for lithostratigraphic correlations (Troelsen 1937; Surlyk 1970b, 1971; Surlyk & Birkelund 1977). Recently, it was proposed to place the Maastrichtian–Danian boundary at the base of the Fish Clay as available evidence suggests the globally isochronous nature of the clay bed (Surlyk 1983).

Facies changes at the Maastrichtian–Danian boundary occur in all the Danish sections. In northwestern Denmark, the boundary is marked by a thin marl layer, but occurs in an otherwise monotonous horizontally bedded sequence of pelagic chalk. This boundary sequence may be somewhat more complete than that at Stevns Klint as suggested by dinoflagellate data (Hansen 1977, 1979b, c). Recently, Nye Kløv was proposed as a substitute for Stevns Klint as stratotype for the Cretaceous–Tertiary boundary, if it can be

demonstrated, that the Stevns Klint section contains a hiatus at the boundary (Surlyk 1983, 1984).

The Maastrichtian chalk of northwestern Europe was deposited in an east–west-running seaway from the Cretaceous Atlantic to Poland (Fig. 2; Håkansson *et al.* 1974). Further eastwards it connected with the wide Russian shelf sea. To the north the Maastrichtian chalk sea was bordered by the Precambrian shield and towards the south by the mid-European Island. There were furthermore connections to the Tethyan sea in both a southeastern and a southwestern direction.

The central facies for the Upper Cretaceous of Northwest Europe is chalk. Many different definitions on the rock term 'chalk' have been proposed (e.g., Scholle 1977; Gealy *et al.* 1971; Bromley 1979; Bromley & Gale 1982) depending on whether there is a need for a lithological or a primary compositional definition. For the purposes of this paper a field definition is used: *chalk* is a partially or completely unlithified white or nearly white pure calcilutite (Bromley 1979). Chalk is composed primarily of calcareous nanno- and microfossils, especially coccoliths, foraminifers, dinoflagellates and calcispheres, all of pelagic origin. Chalk is the central facies and the most common matrix in the Maastrichtian and Danian carbonate facies in Denmark. In addition to the chalk, varying subordinate quantities of sand-sized skeletal material of benthic origin are present. Where skeletons of a particular group dominate the fabric, the subfacies is named after the origin of the skeletal elements, e.g. bryozoan chalk and bryozoan limestone.

Bryozoan limestone is a common Lower Danian facies (e.g. Cheetham 1971; Fig. 3 herein). It contains a higher proportion of bryozoan skeletons than bryozoan chalk though the fabric is still mud supported. Bryozoan limestone is distinctly bedded and in many places deposited in large-scale mound structures such as in the eastern part of Denmark, e.g. Stevns Klint. Many localities in Jylland, e.g. Nye

Kløv, Kjølbygård and Dania, however, exhibit horizontally bedded bryozoan limestone, and these differences in bedding are in all probability due to differences in palaeogeographic position and water depth of the localities. The sequence in Jylland was thus deposited in the axial part of the Late Cretaceous Danish Basin, whereas the Stevns Klint and Karlstrup sections in Sjælland were deposited in shallower water over structural highs (Surlyk 1979b; Figs. 1, 2, 3 herein).

The Nye Kløv locality

The Nye Kløv locality is a small abandoned chalk quarry with average dimensions of 40 m in length and 20 m in height. The quarry is situated in a raised Pleistocene cliff facing the inner drained parts of the Lønnerup Fjord in northern Jylland (Fig. 4). The sequence exposes 8 m of Upper Maastrichtian and 12 m of Lower Danian chalk. The sediments were initially horizontally bedded but appear to have suffered neotectonic movements that tilted the sediments slightly (Håkansson & Hansen 1979).

The sedimentological sequence at the Nye Kløv locality is from bottom to top: (a) Upper Maastrichtian white pelagic chalk with scattered flint nodules and only few macrofossils; (b) 3 cm of brownish-grey marly clay, the base of which marks the Maastrichtian–Danian (Cretaceous–Tertiary) boundary; (c) 0.5 m of Lower Danian greyish marly chalk; (d) 8.5 m white bryozoan limestone containing as much as 25 wt % of bryozoans; (e) 2.0 m of white bryozoan chalk which gradually passes into white pelagic chalk at the top of the section (Figs. 6, 7).

In the Lower Danian the flint is mainly developed as continuous bands. Flint layers appear to reflect the primary bedding in most places, although they do not represent the actual bedding planes, as they are formed by early diagenetic silicification of *Thalassinoides* burrow networks (e.g. Bromley 1975).

The soft, brown boundary clay layer is microconglomeratic and smeared, and contains small angular as well as rounded chalk clasts, reworked from the underlying and more competent Upper Maastrichtian chalk. The clay itself may have acted as a lubrication plane for halokinetically controlled post-sedimentary movements.

The outcrops in northwestern Jylland (Bjerre, Dania, Eerslev, Kjølbygård and Nye Kløv) are situated over salt diapir domes, which probably were not activated until after the Cretaceous–Tertiary deposition (Håkansson & Hansen 1979).

Dinoflagellate data (Hansen 1977, 1979b, c) suggest that there may be a hiatus or at least a lower sedimentation rate between the Upper Maastrichtian and the Lower Danian in some of the localities in eastern Denmark compared to northwestern Denmark. Furthermore, the hiatus, if present, is smaller in the Danish localities than in most other boundary sections reported elsewhere in the world, including deep-sea cores.

Both the uppermost Maastrichtian and the Lower Danian chalk of Nye Kløv is totally burrowed by *Thalassinoides*, *Zoophycos*, *Chondrites* and traces of other organisms. In the Danian the ichnofabric is dominated by *Thalassinoides*,

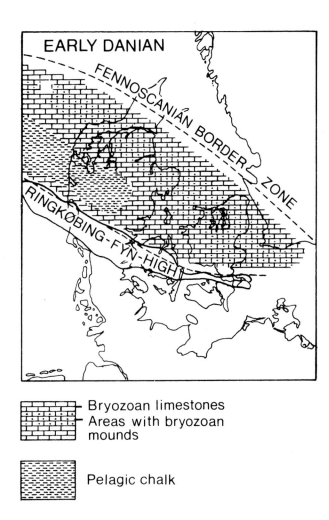

EARLY DANIAN

|☷☷☷☷☷| – Bryozoan limestones
– Areas with bryozoan mounds

|▦▦▦| Pelagic chalk

– – – – – Coastline

Fig. 3. Generalized Early Danian facies distribution (after Håkansson & Thomsen 1979).

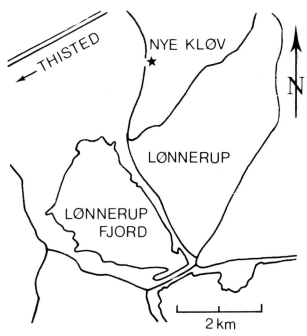

Fig. 4. Map showing position of Nye Kløv facing the inner, drained parts of Lønnerup Fjord in Northern Jylland, Denmark.

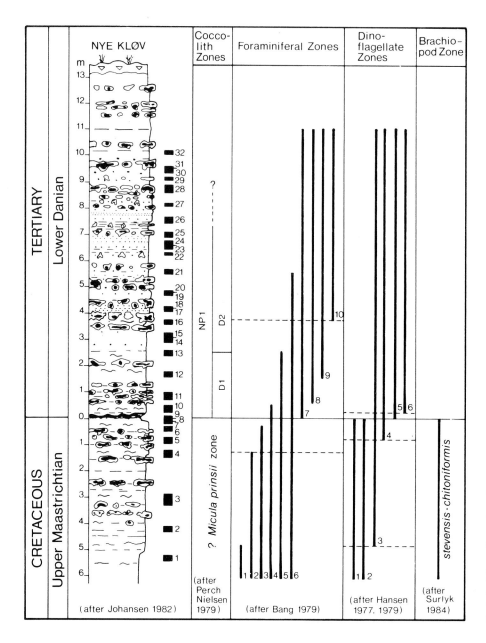

Fig. 5. Compilation of biozones covering the Maastrichtian–Danian boundary interval at Nye Kløv. Coccolith data provided by Perch-Nielsen (1979a, b, c), foraminiferal data by Bang (1979), dinoflagellate data by Hansen (1977, 1979b), and brachiopod data by Surlyk (1972, 1984). Lithological profile measured by Johansen (1982).

Coccoliths: Zone NP 1 (after Martini 1971) is subdivided into the subzones D1 and D2 (after Perch-Nielsen 1979c). Marker species for D1: *Biantholithus sparsus* and for D2: *Zygodiscus sigmoides*.

Foraminifera: (1) *Globotruncana arca*; (2) *Pseudotextularia elegans*; (3) *Globotruncanella petaloides*; (4) Part of the "Lønnerup Assemblage" (Bang 1979); (5) *Heterohelix* spp., *Hedbergella* spp., "*Globigerinella*" *aspera*. (Assemblages 4 and 5 in the Danian are possibly reworked from the Maastrichtian.) (6) *Guembelitria* spp.; (7) *Chiloguembelina* spp.; (8) *Woodringina* sp.; (9) *Eoglobigerina danica* s.l.; (10) *Globoconusa daubjergensis*.

Dinoflagellates: (1) *Spiniferites ramosus cavispinosus*; (2) *Palynodinium grallator*; (3) *Thalassiphora pelagica*; (4) *Chiropteridium* (5) *Danea mutabilis* and *Carpatella cornuta*; (6) *Xenicodinium rugulatum*.

Brachiopods: The *stevensis–chitoniformis* zone of Surlyk (1984), the base of which is defined by the appearance of *Argyrotheca stevensis* and the top by the disappearance of *Magas chitoniformis*.

whereas *Zoophycos* has not been observed. The faunally depauperated lowermost Danian marly chalk is totally bioturbated as well, and *Thalassinoides*, *Planolites* and *Chondrites*-like trace fossils are present here (Ekdale & Bromley 1984).

Stratigraphy

The macrofauna at Nye Kløv is sparse, apart from a significant bryozoan content in the higher part of the Lower Danian (Fig. 7 and Håkansson & Thomsen 1979). The most common macrofossils are *Baculites* and *Inoceramus* from the Upper Maastrichtian, *Echinocorys* and *Terebratulina chrysalis* from the Lower Danian chalk and *Tylocidaris abildgaardi* in the Lower Danian limestone.

The uppermost Campanian and Maastrichtian of northwestern Europe have been biostratigraphically subdivided in great detail on the basis of brachiopods (Bitner & Pisera 1979; Steinich 1965; Surlyk 1969, 1970b, 1972, 1982, 1984). Thus the uppermost Campanian is divided into two brachiopod zones and the Maastrichtian into ten brachiopod zones.

The lower boundary of each zone is defined by Surlyk (1983, 1984) in a stratotype on the first or the last appearance of a species. The upper boundary of each zone is defined by the lower boundary of the succeeding zone. The Upper Maastrichtian of Nye Kløv belongs to the topmost Maastrichtian *stevensis–chitoniformis* Zone (Surlyk 1970b, 1982, 1984; Fig. 5 herein), the top of which is defined by the disappearance of one of the most characteristic Maastrichtian species, *Magas chitoniformis*.

The highest part of the Maastrichtian of Denmark has furthermore been subdivided by means of planktic foraminiferans (Bang 1979), dinoflagellates (Hansen 1977, 1979b, c) and coccoliths (Perch-Nielsen e.g. 1979a, b). Fig. 5 shows the biostratigraphical zonations based on these three groups of micro- and nannofossil and the brachiopod zonations.

Hansen (1979b) stated that the *Chiropteridium inornatum – Palynodinium grallator* dinoflagellate zone is restricted to the topmost 0.5 m of the Upper Maastrichtian and furthermore restricted to the localities deposited in the axial part of the Danish Basin (Bjerre, Dania, Kjølbygård and Nye Kløv). In

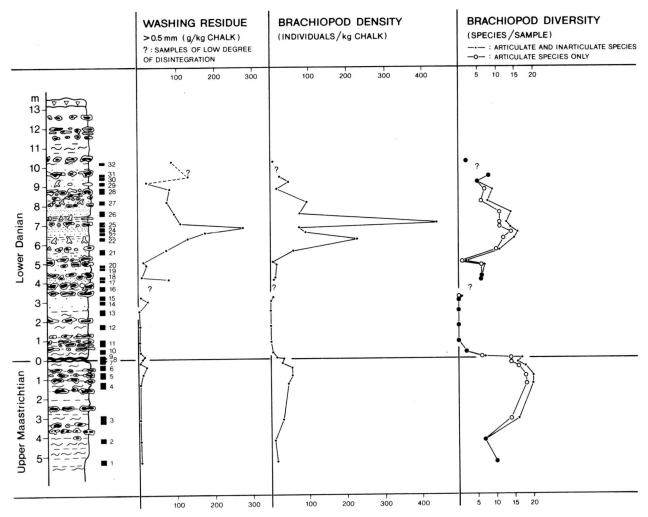

Fig. 6. Diagram showing the density of skeletal fragments (= washing residue), brachiopod density (individuals per kilogram washing residue) and brachiopod diversity (species per sample) in a series of samples across the Cretaceous–Tertiary boundary at Nye Kløv (from Johansen 1982). The diversity is shown both as total number of inarticulate and articulate species and as number of articulate species only.

other localities investigated (Stevns Klint, core Copenhagen Tuba 13), *C. inornatum* occurs only above the Maastrichtian–Danian boundary together with dinoflagellate species characteristic of the Danian, such as *Dania mutabilis* (basal Danian) and *Carpatella cornuta* (Lower Danian). This difference in biostratigraphic development together with lithological evidence suggest a more complete sequence across the Maastrichtian–Danian boundary in northern than in eastern Denmark. It should be mentioned, however, that palaeoenvironmentally controlled differences in the dinoflagellate flora of the Danish Basin do occur, as is reflected in the relative abundance of certain species.

For further discussion of biostratigraphic zonations across the Maastrichtian–Danian boundary references should be made to papers in Birkelund & Bromley (1979) and in Christensen & Birkelund (1979).

Material and methods

Methods

The present work is based on a series of 32 closely spaced samples collected from the Cretaceous–Tertiary boundary

strata from the Nye Kløv locality, northwestern Denmark. The samples cover the topmost 5 m of the Upper Maastrichtian and 13 m of the Lower Danian (Fig. 6). The weight of the samples was in most cases between 5 and 10 kg. A sample weight of 5 kg is considered to be ideal in quantification of the micromorphic brachiopods, as larger samples seldom increase the number of species (Surlyk 1969, 1972). Diversity is thus simply expressed as the number of species per chalk sample. A few of the samples weighed less than 5 kg, however, and for reasons of standardisation all numbers of density and washing residue are recalculated to a sample weight of 1 kg. The samples were washed with a method involving repeated freezing and heating in a supersaturated Glauber-Salt solution, as described in detail by Surlyk (1972). The process was repeated 10–18 times. After treatment, the chalk breaks down completely into mud and very clean fossils. The samples were then washed through a 0.25 mm sieve, the washing residue dried slowly and finally hand-sieved into 0.25–0.50 mm, 0.5–1.0 mm and >1.0 mm fractions. The brachiopods were picked from the 0.5–1 mm and >1 mm fractions under a binocular microscope at ×6, ×12 and ×25 magnification. The number of individuals was estimated by adding the number of complete shells to either the number of dorsal or ventral valves depending on the

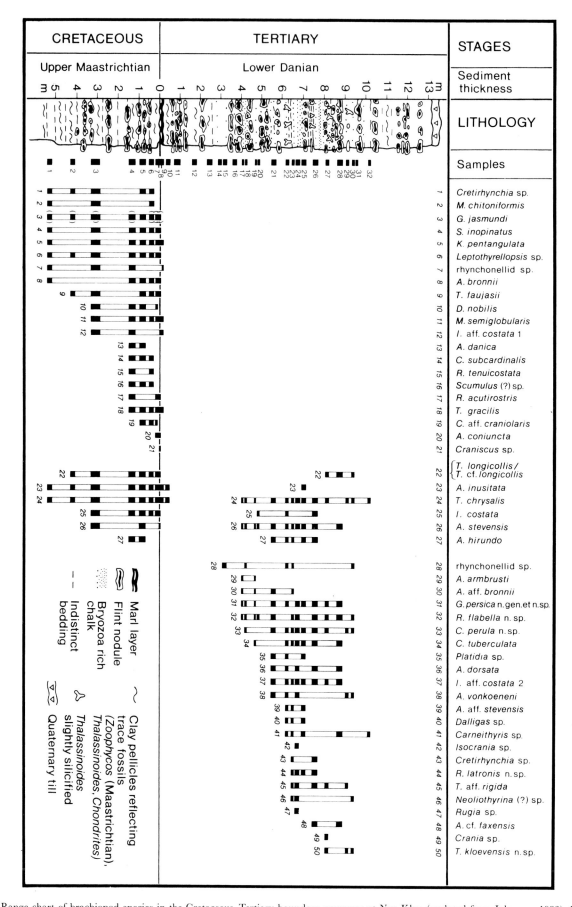

Fig. 7. Range chart of brachiopod species in the Cretaceous–Tertiary boundary sequence at Nye Kløv (updated from Johansen 1982). NK7, NK8 and NK9 are separate samples, but as these were very closely taken, slight overlap occurs. Sample NK7 thus contains the topmost Maastrichtian chalk and is in contact with the boundary clay. Sample NK8 is a sample of the 3 cm thick Maastrichtian–Danian boundary clay and may contain a few centimetres of the succeeding Lower Danian marl. Sample NK9 contains the lowermost Danian marl deposited immediately after the boundary clay and may contain the top of the latter. The horizontal bars between the samples NK7, NK8 and NK9 show the vertical extension of these samples.

greater number. In addition, the minimum number of individuals represented by fragments was estimated.

The figured specimens were prepared with an ultrasonic cleaner before Scanning Electron Microscope photography using either a Stereoscan 180 or a Philips 515 at the Institute of Historical Geology and Palaeontology, Copenhagen. In order to describe the important internal morphological elements of the brachiopods, a number of complete shells have been opened by application of the technique described by Surlyk (1969).

After taxonomic determination of the fossils to species level quantitative graphs were constructed (Figs. 6, 7, 30–34).

Material

About 3000 brachiopod specimens were studied from Nye Kløv and in addition a large number of specimens were examined from Coniacian–Maastrichtian strata of Northwest Germany, Campanian–Maastrichtian strata in Norfolk, England, from Upper Maastrichtian–Danian strata at Limburg, Holland and Belgium, and from Upper Maastrichtian–Danian strata in Denmark. The Danish localities are Bjerre, Eerslev, Kjølbygaard, Dania, Rørdal and to a minor extent Gravlev, Nyvanggård, Skellingbro and Vokslev, all in the western part of Denmark, and Karlstrup, Stevns Klint, Fakse and Møns Klint in eastern Denmark. All of the Danish localities except Rørdal, Fakse and Møns Klint expose the Maastrichtian–Danian boundary (Fig. 1). In addition, brachiopod collections from Natuurhistorisch Museum, Maastricht, the Netherlands, Geological Museum, University of Copenhagen, Institute of Palaeoecology, Århus, Denmark and Sedgwick Museum, Cambridge, England have been at my disposal in the course of this study. The type material is housed at the Geological Museum, University of Copenhagen, and is prefixed MGUH.

Preservation

In general, the brachiopods from the Upper Maastrichtian are very well preserved. Preservation of the Lower Danian specimens is more variable, mainly due to a higher degree of calcite overgrowth. In some levels the brachiopods are clean and well preserved and in others the specimens are covered with a thin cemented layer of coccolith ooze. Material from the Lower Danian is generally more recrystallised, and the degree of silicification in the Lower Danian is, at certain levels, so advanced that disintegration of samples is impossible.

Brachiopod terminology, measurements and morphology

The taxonomy follows the *Treatise on Invertebrate Paleontology (H) Brachiopoda* (Moore 1965), and the terminology follows that used by Williams & Rowell (1965). The measured and described characters of the brachiopods in this paper are shown in Figs. 8–11.

Ratios of morphological dimensions are calculated and plotted into scatter diagrams to show the degree of variation

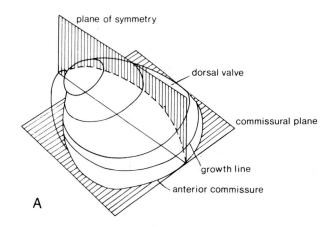

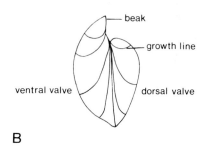

Fig. 8. □A. Plane of symmetry and commissure on a stylized articulate brachiopod in oblique dorso-ventral view. □ B. A stylized articulate brachiopod in lateral view.

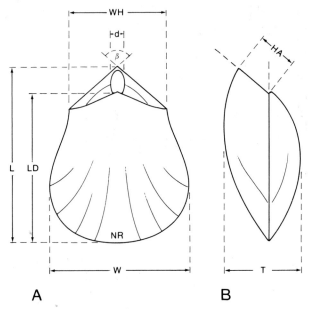

Fig. 9. Measured morphological characters on a typical, *Terebratulina*-like brachiopod. L: length, LD: length of dorsal valve, W: width, T: thickness, d: width of foramen, WH: width of hinge line, HA: heigth of area, ß: angle of area edge (°), NR: number of ribs.

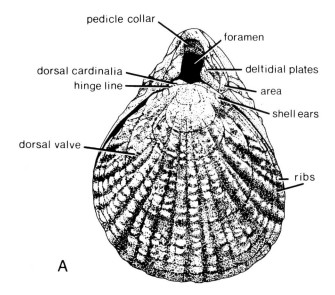

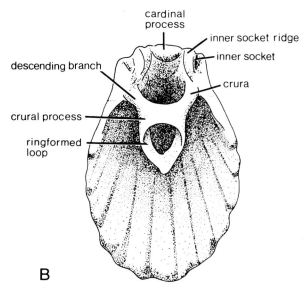

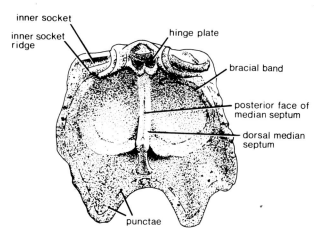

Fig. 10. □ A. Morphological elements of the exterior of *Terebratulina* aff. *rigida* shown in dorsal view. □ B. Morphological elements of the interior of a dorsal valve of *Terebratulina longicollis* showing the cardinalia and the brachidium.

Fig. 11. Morphological elements of the interior of a dorsal valve of *Argyrotheca hirundo.*

within the species (Figs. 12–29). Such ratios are plotted for selected species on published scatter diagrams of other authors for comparison. For this purpose scatter diagrams from Steinich (1965, 1968a, b), Surlyk (1969), and Bitner & Pisera (1979) have been used.

The following morphological dimensions are illustrated: length to width (L/W); length of dorsal valve to width (LD/W); thickness to width (T/W); width of foramen to width (d/W); width of hinge line to width (WH/W); and number of ribs to width (NR/W). Area edge angle (beta in degrees) was measured for a single species (*Scumulus inopinatus*).

Systematic descriptions

Order Rhynchonellida Kuhn 1949

Family Rhynchonellidae Gray 1848

Genus indet.

Material. – Three complete shells, 8 dorsal and 2 ventral valves and a large number (100–150) of fragments. The material is considered to belong to the family Rhynchonellidae because of the characteristic glossy fracture pattern reflecting the fibrous shell structure. The shells are furthermore thick and impunctate, but the available specimens do not allow further classification.

Occurrence. – The material occurs in the Upper Maastrichtian samples NK1, NK3, NK4, and NK9, and in the Lower Danian samples NK15, NK18, NK22, NK23 and NK30.

Subfamily Cyclothyridinae Makridin 1955

Genus *Cretirhynchia* Pettitt 1950

Type species. – *Terebratula plicatilis* J. Sowerby 1816, by original designation.

Cretirhynchia sp.
Pls. 1:1–2; 2:1

Synonymy. – □ 1984 *Cretirhynchia* sp. – Surlyk & Johansen, Fig. 1.

Material. – The material consists of 14 complete shells, 15 dorsal and 9 ventral valves. All but one of the complete shells are juveniles and thus the material does not allow for any adequate description of ontogenetic development.

Description. – The adult shell is a few centimetres in length and smooth. The outline is in the juvenile stages pointed oval; maximum width changes during ontogeny from about the midline of the shell to the anterior shell margin. The shell is planoconvex in the juvenile stages, and the shell ears very indistinct. The beak is suberect to erect in the juvenile stages and during ontogeny becomes slightly incurved. The foramen is large and triangular in juveniles. Later the deltidial plates restrict the foramen completely anteriorly. Hence the exterior opening of the foramen is small, circular and submesothyridid in adults. The deltidial plates in the juveniles have a characteristic growth pattern which results in two laterally directed wing-like doublings of the deltidial plates. This pattern is a result of changes in growth direction of the deltidial plates during ontogeny (Steinich 1965, p. 21–22, Fig. 5). The characteristic deltidial plates distinguish the rhynchonellid juveniles from other smooth-shelled forms. The hinge is strong and well developed. In the juvenile stages, the inner socket ridges are low, short and strongly converging anteriorly. The inner socket ridges continue anteriorly into the crura, which in juveniles are strong, long and ventrally directed. The teeth are blunt and sharp, and no cardinal process is seen. In larger forms, a low median furrow is seen in the floor of the dorsal valve. The shell is rather thick.

Remarks. – Only two rhynchonellid species have been described from the Danish Maastrichtian chalk: *Cretirhynchia retracta* (Roemer 1841) and *Cretirhynchia limbata* (Schlottheim 1813) (Surlyk 1972).

The two species are very closely related and very difficult to distinguish from each other, particularly in the case of juveniles. The Lower Danian specimens are also juveniles, and there seems to be no marked difference between these specimens and the juveniles of the Maastrichtian species. Neither *C. retracta* nor *C. limbata*, however, have been previously recorded from the Danian.

Occurrence. – Most of the material occurs in the Upper Maastrichtian samples NK1, NK2, NK3, NK5, and NK6. The Lower Danian specimens occur in samples NK23 and NK26.

Family Cryptoporidae Muir-Wood 1955

Genus *Cryptopora* Jeffreys 1869

Type species. – *Atretia gnomon* Jeffreys 1876, by original designation (monotypy).

Cryptopora perula n.sp.
Pl. 3:1–7; Figs. 12A–D, 13A–C

Synonymy. – □ 1984 *Cryptopora* n.sp. – Surlyk & Johansen, Fig. 1.

Derivation of name. – Latin *perula*, pearl; referring to the droplike shape of the shell.

Holotype. – MGUH 16906, Pl. 3:1A, B; Lower Danian, Nye Kløv, sample NK24. Type locality: Nye Kløv, north of Thisted, northwestern Denmark. Type horizon: Lower Danian.

Material. – 165 complete shells, 63 dorsal and 76 ventral valves. The largest specimen is from the Lower Danian sample NK25 and has the following dimensions: Length 2.44 mm; dorsal valve length 2.00 mm; width 1.92 mm;

thickness 0.64 mm; foramen width 0.40 mm. Measurements of the holotype: Length: 1.88 mm; dorsal valve length: 1.48 mm; width: 1.40 mm; thickness: 0.52 mm; foramen width: 0.28 mm.

Diagnosis. – Shell small, flat biconvex, impunctate, relatively thin, and smooth with subtriangular outline. Foramen large, hypothyridid and triangular. Deltidial plates indistinct, disjunct. Floor of ventral valve contains thickened area wedging out towards apex. Median septum high, narrow, short pointed, crura long, slender and ventrally directed, rising from base of socket ridges.

Description. – The shell is small with a rounded oval to subpentangular outline in juveniles, whilst adults are subtriangular to pointed oval. In juveniles the maximum width is at the anterior margin of the shell, but during ontogeny draws backwards to around midway shell length. The ratios of L/W, LD/W, T/W and d/W are shown in Fig. 12A–D. The auricles are indistinct, and the shell flattened biconvex. Some large specimens are resupinate. The shell surface is smooth, and growth lines are only very rarely observed. The hinge line is very short and straight. The beak is relatively short, nearly straight in the juvenile stages and suberect to erect later in ontogeny. Both the area and deltidal plates are indistinct, and in most specimen the foramen is very wide, consequently leaving very little space for deltidial plates and area. The deltidial plates are narrow, low ridges that limit the foramen laterally. The foramen is hypothyridid, large and broad to pointed triangular, and there is no pedicle collar. The teeth are short and pointed; they converge dorsally and seem to form an anterior projection of the deltidial plates, and there are no dental plates. The posterior median area of the interior of the ventral valve is thickened; this elevated area wedges out towards the apex. The inner socket ridges are equally low but distinct, and they converge sharply posteriorly. The cardinal process is small but marked and the crura rise from the base of the socket ridges converging rapidly midventrally. The descending arms of the brachidium are long and slender. In larger forms two small dorsally converging crural processes occur where the crura meet the descending branches. The latter reach the median septum level at a shell length of 1.25 mm. The median septum is short, high, narrow and pointed. The posterior edge is almost perpendicular to the shell floor, and the anterior edge slopes steeply towards the shell margin. The median septum rises around midway shell length and retains this position during ontogeny. The lophophore was an early spirolophe to possibly a fully developed spirolophe. The spicular skeleton was heavily spiculated, and in several specimens a recrystalised but fully preserved skeleton can be seen. The shell is relatively thin.

Remarks. – The genus *Cryptopora* Jeffreys has hitherto been reported only from the Eocene to Recent (Muir-Wood 1965; Cooper 1979); and the genus has been described and well illustrated by several authors (e.g., Davidson 1874; Cooper 1959, 1971, 1973b, 1979).

Bitner & Pisera (1979) established the genus *Cryptoporella*. The name refers to the close affinity with *Cryptopora*. *Cryptoporella* occurs in Upper Campanian–Lower Maastrichtian

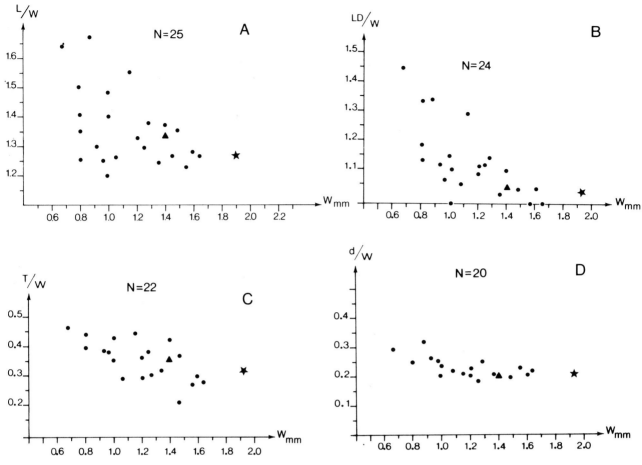

Fig. 12. Scatter diagrams of *Cryptopora perula* n.sp. from the Lower Danian, Nye Kløv. ▲: holotype, sample NK24, Lower Danian, MGUH 16906; ★: largest specimen, sample NK25, Lower Danian. □A. Ratio shell length L (mm) to width W (mm). □B. Ratio dorsal valve length LD (mm) to width W (mm). □C. Ratio thickness T (mm) to width W (mm). □D. Ratio foramen width d (mm) to width W (mm).

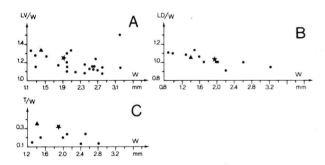

Fig. 13. Scatter diagrams of *Cryptoporella antiqua* Bitner & Pisera (●), Upper Campanian – Lower Maastrichtian, Mielnik, Poland (from Bitner & Pisera 1979, Fig. 3a–c), with holotype (▲) and largest specimen (★) of *Cryptopora perula* n.sp. inserted for comparison (cf. Fig. 12). LV: ventral valve length (mm), LD: dorsal valve length (mm), T: thickness (mm), W: width (mm).

chalk of the Mielnik Chalk Pit, Eastern Poland. *Cryptoporella* lacks, however, a median septum, the crura are very short, and the foramen is much wider than in *Cryptopora*.

The *Cryptopora* material from Nye Kløv has affinities to both genera. The foramen and the hinge are reminiscent of *Cryptoporella*, whereas the crura and the median septum are formed as in *Cryptopora*. As the median septum and the crura appear to represent the most important taxonomic characters, it is preferred to refer the material to *Cryptopora* rather than to *Cryptoporella*. Of known *Cryptopora* species, *Cryptopora*

perula is most reminiscent of *Cryptopora gnomon* Jeffrup. The latter differs from other species of *Cryptopora* sp. in possessing non-alate deltidial plates (Cooper 1979). Fig. 13 shows L/W, LD/W and T/W ratios for holotype and largest specimen of *Cryptopora perula* for comparison with *Cryptoporella antiqua* Bitner & Pisera (1979).

Occurrence. – *Cryptopora perula* first appears 5 m above the Maastrichtian–Danian boundary and occurs in most Lower Danian samples above this level (NK18, NK21, NK22, NK23, NK24, NK25, NK26, NK27, NK29, NK30). It varies in numbers from a few to many individuals and is the dominating species in a few samples (NK22, NK25). This is the first recorded occurrence of the species in the Lower Danian.

Order Terebratulida Waagen 1883

Family Terebratulidae Gray 1840

Subfamily Rectithyridinae Muir-Wood 1965

Genus *Neoliothyrina* Sahni 1925

Type species. – *Terebratula obesa* Davidson 1852, by original designation. *Neoliothyrina obesa* Sahni 1925 is considered a junior synonym.

Neoliothyrina? sp.
Pl. 1:5, 6

Synonymy. – □ 1984 *Neoliothyrina*(?) – Surlyk & Johansen, Fig. 1.

Material. – Three fragmented dorsal valves and two broken beaks from the Lower Danian. No measurements have been obtained.

Description. – The material does not allow for any adequate description, but judged from the broken material, the adult shell appears to be large, flattened biconvex, with a short, erect to slightly incurved and strongly worn beak. The reference to *Neoliothyrina?* is, however, based mainly on the characteristic shape of the inner socket ridges. The hinge line is short, and the inner socket ridges are short, low, parallel and laterally flattened. The posterior margins of the inner socket ridges constitute, with the cardinal process, a low horizontal bridge. The cardinal process is relatively large and bulbous. Anterior to the cardinal process, two concave, anteriorly positioned elongated plates represent the hinge plates. The anterior margin of the hinge plates continues into the crural bases. The shell is thin.

Remarks. – The hinge area of the Upper Cretaceous *Neoliothyrina obesa* Sahni 1925 differs from *Neoliothyrina?* sp. in having markedly diverging inner socket ridges and very broad hinge plates. The species found is reminiscent in the hinge area of the Upper Cretaceous *Neoliothyrina fittoni* (Hagenow) (Asgaard 1972), but no further statement can be obtained.

Occurrence. – The species is found in samples NK23, NK24 and NK30 from the Lower Danian. A number of relatively large, thin-shelled, punctate and smooth fragments are found in sample NK18. These may belong to *Neoliothyrina?* sp.

Subfamily Carneithyridinae Muir-Wood 1965

Genus *Carneithyris* Sahni 1925

Type species. – *Carneithyris subpentagonalis* Sahni 1925, by original designation.

Carneithyris subcardinalis (Sahni 1925)
Pl. 2:2

Synonymy. – □ 1842 *T. carnea* Sow. – Hagenow, p. 539, no. 13. □ 1894 *Terebratula carnea* Sowerby – Posselt, p. 3 , no. 29. □ 1909 *Terebratula carnea* Sowerby – Nielsen, p. 163, Pl. 2:68–77. □ 1925 *Chatwinothyris subcardinalis* sp.n. – Sahni, Fig. 9; Pl. 14:4, 4a. □ 1963a *Chatwinothyris subcardinalis* Sahni – Steinich, p. 608, Fig. 9. □ 1965 *Chatwinothyris subcardinalis* Sahni – Steinich, pp. 37–46, Figs. 24–34, Pl. 5:4, Pl. 6:1a–d, 2a–d, 3, 4, Pl. 7:1a–b, 2, Pl. 11:1, 2a–b, 3. □ 1970 *Carneithyris* sp. Sahni – Asgaard, pp. 361–367. □ 1972 *Carneithyris subcardinalis* (Sahni) – Surlyk, p. 24, Figs. 5, 11, 16, 17, 18, Pl. 5C. □ 1975 *Carneithyris subcardinalis* (Sahni) – Asgaard, pp. 320–232, 335–339, 360, Pl. 8:1–4. □ 1979 *Chatwinothyris subcardin-*

alis Sahni – Bitner & Pisera, p. 73, Pl. 2:5–6. □ 1982 *Carneithyris subcardinalis* (Sahni) – Surlyk, Fig. 1, Pl. 1b. □ 1984 *Carneithyris subcardinalis* (Sahni) – Surlyk & Johansen, Fig. 1.

Material. – Fifteen complete shells and two ventral valves. All the identified specimens are juveniles less than 2.0 mm in width, and no measurements have been obtained. A number of fragments most likely belonging to larger specimens of *C. subcardinalis* are also present.

Description. – The juvenile shells are smooth, thin, containing large punctae, and are very elongated, pointed oval in outline. Larger specimens are also smooth- and thin-shelled, broad oval to subrhombic in outline. The adult specimens are of centimetre-size, as inferred from the size of fragments. The juvenile shell is biconvex with maximum width at the anterior shell margin. The anterior commissure is straight, the hinge line is short, weakly oblique and the auricles are indistct. The beak is low, acute and erect to slightly incurved, and the foramen is very small, subtriangular and hypothyridid in these young specimens. The foramen is laterally and anteriorly delimited by two large, smooth, triangular, plate-like, deltidial plates. The hinge is poorly developed, the inner socket ridges are low, the hinge teeth are small, and no cardinal process is present.

Remarks. – The material is referred to *Carneithyris subcardinalis* based on the characteristic small foramen and large deltidial plates, the very thin shell and poorly developed hinge of the juveniles. Considerable change is taking place during growth, but the material does not allow any adequate description of this. For more detailed descriptions of the ontogenetic developement see Steinich (1965), Surlyk (1969) and Asgaard (1970, 1975).

Occurrence. – *Carneithyris subcardinalis* was found in the Upper Maastrichtian samples NK4, NK5, and NK6. In Denmark the species becomes extinct at the top of the Maastrichtian and does not occur in the Lower Danian.

Carneithyris sp.
Pl. 1:2–4

Synonymy. – □ 1984 *Carneithyris*(?) sp. – Surlyk & Johansen, Fig. 1.

Material. – Ten complete shells, 3 dorsal and 6 ventral valves and a number of fragments. All the complete specimens are juveniles, and no measurements have been obtained on adult forms.

Description. – The outline is very elongated and pointed oval, and the largest complete specimen has a shell width of 1.50 mm and length of 1.80 mm. The maximum width lies at the anterior shell margin, and the auricles are indistinct. The shell is biconvex, with a very convex ventral valve, and the anterior commissure is straight. The shell is smooth. There are a few distinct growth lines. The hinge line is very short and oblique laterally, and the beak is erect to slightly incurved, very short and blunt. The area is indistinct and narrow, and the foramen is very small, triangular and hy-

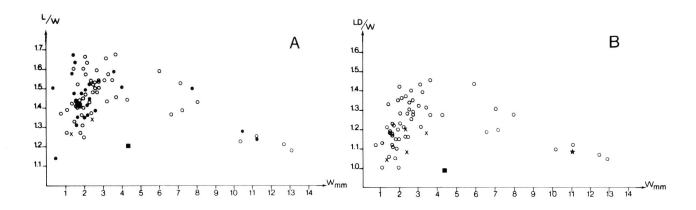

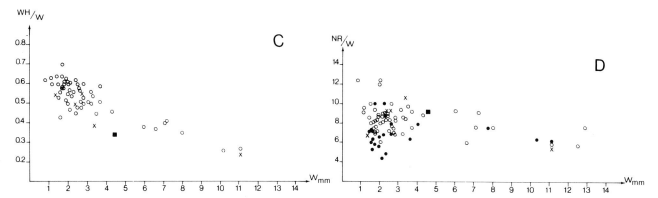

Fig. 14. Scatter diagrams of *Terebratulina chrysalis* (Schlottheim) from the Lower Danian, Nye Kløv (×; ★: largest specimen, sample NK30) and from the Lower Maastrichtian, Rügen (○, ●; data from Steinich 1965, Figs. 44, 45, 47, 51), and of *T.* aff. *rigida* from the Lower Danian, Nye Kløv (■: largest specimen, sample NK29). □A. Ratio shell length L (mm) to width W (mm). □B. Ratio dorsal valve length LD (mm) to width W (mm). □C. Ratio hinge line width WH (mm) to width W (mm). □D. Ratio number of ribs NR to width W (mm).

pothyridid, considerably limited laterally by broad, triangular deltidial plates. The deltidial plates meet posteriorly, and a short pedicle collar is developed. The basis of the crura develop as long, converging ridges close to the inner surface of the dorsal valve, anterior to the short, low inner socket ridges. The teeth are small and blunt, and the shell is relatively thick and punctate.

Remarks. – The species is referred to the genus *Carneithyris*, due to the very small foramen limited by the plate-like triangular deltidial plates and to the very elongated pointed oval outline of the juvenile shell. The genus has already been recognized from the Danian of Denmark and Sweden by Asgaard (1975), but the present material does not allow for any statement on the adult forms.

Occurrence. – The species is found in the Lower Danian samples NK22, NK23, NK25, NK28, and NK32.

Family Cancellothyrididae Thomson 1926

Subfamily Cancellothyridinae Thomson 1926

Genus *Terebratulina* d'Orbigny 1847

Type species. – *Anomia caputserpentis* Linné 1767, by original designation. *Anomia retusa* Linné 1758 is considered a junior synonym.

Terebratulina chrysalis (Schlottheim 1813)
Pl. 4:1–5; Fig. 14A–D

Synonymy. – □ 1813 *Terebratulites chrysalis* Schlottheim, p. 113 (cit. Faujas 1798, Pl. 26:7, 9). □ 1828 *Terebratula striatula* Mant. – Beck, p. 581. □ 1894 *Terebratulina striata* Wahlenberg – Posselt, p. 32, no. 19, p. 5, footnote 3; p. 10. □ 1903 *Terebratulina striata* Wahlbg.sp. – Ravn, p. 377, no. 18, pp. 390–391, no. 18, pp. 414–415, no. 18. □ 1909 *Terebratulina striata* Wahlenberg – Nielsen, p. 159, no. 20, Pl. I:28–32, p. 134, no. 20, p. 137, no. 20, pp. 138–139, no. 20, p. 141, no. 20, p. 144. □ 1953 *Terebratulina chrysalis* (von Schlottheim) – Wind, p. 79. □ 1965 *Terebratulina chrysalis* (Schlottheim) – Steinich, pp. 53–66, Figs. 44–61. Pl. 8:1a–d. Pl. 9:1–5, 9a, b, 10a, b. □ 1968 *Terebratulina chrysalis* (Schlottheim) – Popiel-Barczyk, pp. 63–65, Fig. 25, Pl. 17:1–3. □ 1972 *Terebratulina chrysalis* (Schlottheim) – Surlyk, pp. 21, 22, 23, Figs. 5, 12, 13, 14, 15, 16, 17, 18, Pl. 4:b, h. □ 1977 *Bisulcina chrysalis* (Schlottheim) – Titova, pp. 81–82, Fig. 6. Pl. 10:1. □ 1979 *Terebratulina chrysalis* (Schlottheim) – Bitner & Pisera, pp. 73–74, Pl. 3:12–15. □ 1984 *Terebratulina chrysalis* (Schlottheim) – Surlyk & Johansen, Fig. 1.

Material. – The material is very fragmented, and the majority of the specimens are juveniles that may be confused with juveniles of other species. However, the material includes at least 50–60 complete specimens, 60–70 dorsal and 60–70 ventral valves, and a large number of fragments. Complete shells are juvenile forms, but a number of fragments with as many as 50 ribs are found and belong to very large specimens. The largest measured specimen is represented by a dorsal valve from the Lower Danian sample NK30 and has the following dimensions: Length 12 mm; width 11 mm; hinge line width 3 mm; number of ribs 60.

Description. – The shell is relatively large and thin, and the outline changes during growth from elongated subtriangular to elongated subpentagonal. The maximum width lies near the mid-length of the shell. The auricles are well defined limited and large in juveniles, while they are small and oblique in large forms. The shell is flattened biconvex, and the anterior commissure is rectimarginate to weakly sulcate. The shell surface posseses 8–>60 straight ribs, and new ribs are formed by intercalation. The ribs are straight, strong and coarsely knobbed in the juveniles. The rib sculpture consists of knots placed on top of the ribs, but never forms transverse half-rings as are seen in *T. faujasii* and *T. longicollis*. The rib width does not change during growth, and the interspace between the ribs is very wide in large forms. In large forms the last formed ribs are very narrow and only weakly sculptured, and only a few distinct growth lines are seen. The beak is short and suberect, and the area is narrow and distinct. The foramen changes from subtriangular and hypothyridid in juveniles, through oval and submesothyridid in intermediate sizes, to circular and mesothyridid in large forms. The deltidial plates are triangular and limit the foramen anteriorly. The hinge is relatively weak, with thin inner socket ridges and small, pointed teeth. A low cardinal process is observed in large forms. The brachidium consists of short, heavy and ventrally converging descending branches that fuse into a broad, transverse bridge. The crural processes fuse mid-ventrally into a broad, transverse bridge, the descending branches and the crural processes together forming a closed tube-shaped brachial ring. A collar representing the anterior part of the spicular skeleton lies close to the dorsal part of the brachial ring and the crura (Pl. 4:2).

Remarks. – *Terebratulina chrysalis* is distinguished from other ribbed cancellothyridid brachiopods by the characteristic rib pattern. The juveniles of *T. chrysalis* are characterised by the early formed large number of ribs, and by possessing large shell ears.

The specimens from Nye Kløv are consistent with descriptions of *T. chrysalis* by, among others, Steinich (1965) and Surlyk (1972). The fossil *T. chrysalis* has two very closely related recent relatives, *T. septemtrionalis* Couthoy and *T. retusa* (Linnaeus), and the three species differ mainly in their spicular skeletons and only in minor differences in the rib patterns.

The morphological similarities between the Maastrichtian and the Lower Danian forms are great, and the Lower Danian specimens are also referred to *Terebratulina chrysalis*. For comparison the Lower Danian material is shown in morphological scatter diagrams of *T. chrysalis* from the Maastrichtian of Rügen (Fig. 14A–D).

Occurrence. – The species occurs in the Upper Maastrichtian samples NK1, NK2, NK3, NK4, NK5, NK6, NK7, NK8, NK9; and the Lower Danian samples NK10, NK17, NK18, NK19, NK21, NK22, NK23, NK24, NK25, NK26, NK27, NK29, NK30, NK32.

Terebratulina faujasii (Roemer 1841)
Pl. 7:3

Synonymy. – ☐ 1841 *Terebratula faujasii* N. – Roemer, p. 40, no. 24, Pl. 7:8. ☐ 1894 *Terebratulina striata* Wahlenberg – Posselt, p. 32, no. 19. ☐ 1963a *Terebratulina faujasii* (Roemer) – Steinich, p. 609, Fig. 7. ☐ 1965 *Terebratulina faujasii* (Roemer) – Steinich, pp. 72–81, Figs. 76–94, Pl. 9:6–8, Pl. 10:1a–d, 2–5. ☐ 1968 *Terebratulina faujasii* (Roemer) – Popiel-Barczyk, pp. 66–67, Pl. 17:4a–b, non Pl. 17:5a–b. ☐ 1972 *Terebratulina faujasii* (Roemer) – Surlyk, pp. 18, 19, 34, 35, 37, Figs. 5, 12, 14, 15, 16, 18. Pl. 2:c–g. ☐ 1979 *Terebratulina faujasii* (Roemer) – Bitner & Pisera, pp. 74–75. Pl. 2:1–3. ☐ 1982 *Terebratulina faujasii* (Roemer) – Surlyk, Fig. 1, Pl. 1c, d. ☐ 1984 *Terebratulina faujasii* (Roemer) – Surlyk & Johansen, Fig. 1.

Material. – Nineteen complete shells, 10 dorsal valves, 11 ventral valves, and a number of fragments. The largest specimen is represented by a dorsal valve from the Upper Maastrichtian sample NK2 and has the following dimensions: length 3.6 mm; width 2.8 mm; number of ribs 10.

Description. – The shell is relatively small; the outline changes from broad subtriangular in the juveniles to elongated subtriangular in the later stages. The maximum width always lies at the anterior shell margin. The shell is biconvex with a highly convex ventral valve. The auricles are large, well developed, and the front commissure is straight. The shell surface possesses 8–10 straight, coarse singular ribs. At a shell width of around 0.4 mm the final number of ribs is formed. The rib sculpture consists of very coarse transverse half rings on top of the ribs. Due to the coarse rib sculpture, it is difficult to trace distinct growth lines. The hinge line is broad, weakly oblique and the beak is rather low, acute and suberect in the juveniles through suberect to erect, and blunt, later on. The area is narrow, triangular and well defined. The foramen is large, subtriangular and hypothyridid in juveniles; in the adults, the foramen becomes submeso-mesothyridid. The foramen is laterally delimited by deltidial plates forming well developed triangular plates. A pedicle collar is well developed. The inner socket ridges are high and laterally concave, and the outer socket ridges are low but distinct. The cardinal process is relatively large and flat, and the hinge teeth are large, triangular and acute. The brachidium is composed of strong crura developed anteriorly to the inner socket ridges, and of descending and ascending branches. The descending branches are fused dorsally into a narrow bridge, and the crural processes ventrally into an equally narrow bridge, thereby forming the closed brachial ring so characteristic for the genus *Terebratulina*. The distance between the descending branches is relatively high in *T. faujasii*. The spicular skeleton was heavy and the lophophore was a plectolophe. The inner surface of the shell is a negative relief of the outer surface and the shell rather thick.

Remarks. – *Terebratulina faujasii* differs from other species of

Terebratulina in its broad subtriangular outline, in possessing a rib sculpture consisting of coarse transverse half rings, in its singular ribs, and in its narrow brachial ring.

The observed species is consistent with the descriptions of *T. faujasii* by Steinich (1965).

Occurrence. – *Terebratulina faujasii* occurs in the Upper Maastrichtian samples NK2, NK3, NK5, NK6, NK7, and NK8. The number of individuals is always low at Nye Kløv. The species becomes extinct at the Maastrichtian–Danian boundary.

Terebratulina gracilis (Schlottheim 1813)
Pl. 7:7–9

Synonymy. – □ 1813 *Terebratulitis graciles* – Schlottheim, p. 113, Pl. 3:3a, b. □ 1852 *Terebratula gracilis* Schlottheim – Davidson, Part 2, p. 38, no. A, Pl. 2:14. □ 1894 *Terebratulina gracilis* Schlottheim – Posselt, p. 33, no. 20, Pl. 3:5–7. □ 1903 *Terebratulina gracilis* v. Schlottheim sp. – Ravn, p. no. 19, pp. 390, 414. □ 1909 *Terebratulina gracilis* Schlottheim – Nielsen, p. 161, no. 22, p. 134, no. 22, p. 137, no. 22. □ 1963a *Terebratulina gracilis* (Schlottheim) – Steinich, p. 605, Figs. 1–3, p. 606, Fig. 4, p. 607, Fig., 5. □ 1965 *Terebratulina gracilis* (Schlottheim) – Steinich, pp. 81–93, Figs. 95–114. Pl. 11:1a–d, Pl. 12:1, 2a, b, Pl. 13:1–3. □ 1972 *Terebratulina gracilis* (Schlottheim) – Surlyk, pp. 24, 40, 43, 44, Figs. 5, 9, 11e, 13, 15b, 18, 19, 20, 21, Pl. 5C. □ 1984 *Terebratulina gracilis* (Schlottheim) – Surlyk & Johansen, Fig. 1.

Material. – Eight complete shells, 26 dorsal and 11 ventral valves and a number of fragments. The largest specimen is from the Upper Maastrichtian sample NK5 and has the following dimensions: length 6.4 mm; dorsal valve length 5.6 mm; width 5.6 mm; thickness 2.2 mm; foramen width 0.2 mm; number of ribs 40.

Description. – The material is consistent with the descriptions of *Terebratulina gracilis* (Schlottheim 1813) given by, among others, Steinich (1965) and Surlyk (1972). Steinich (1965, pp. 81–82) gives a more complete list of synonyms.

The species is rather large, thick-shelled. The outline is elongated oval to subcircular and the maximum width is near the mid-length of the shell. The auricles are very small. The shell is plano-convex to convavo-convex and the front commissure is straight to uniplicate. In very early growth stages, 8–10 straight, finely knobbed, strong radial ribs are present. During ontogeny the number of ribs increases by intercalation to 50–60, and in larger forms the ribs are reflexed towards the lateral margins of the shell. Numerous growth lines are seen. The hinge line is very short, oblique, and continues gradually into the lateral shell margins. The beak is short, acute and changing from erect to incurved during growth. No area is present in large forms. The foramen is very small, oval, submesothyridid to permesothyridid, and restricted laterally and anteriorly by triangular deltidial plates. The pedicle collar is well developed and may in large forms constitute a hollow cylinder. The hinge is well developed and strong, and the brachidium consists of rather thin crura and a closed subtriangular brachial ring. The midventral part of the brachial ring is strongly concave

ventrally and a posteriorly directed, ventrally arching process with two deep lateral incisions is present. The dorsal part of the brachial ring is a wide, slightly arched bridge. The lophophore was a plectolophe. The inner surface of shell is smooth.

Remarks. – *Terebratulina gracilis* differs from other species of *Terebratulina* in its planoconvex to convavo-convex shape, in the short, incurved beak, in its hinge line continuing gradually into the lateral shell margins, in its very small foramen, in exhibiting no area, and in its high number of reflexed ribs. *T.* aff. *rigida* (Sowerby) differs from *T. gracilis* in being of smaller size, in possesing a less incurved beak and a larger foramen, and in the ribs being much less reflexed.

Occurrence. – *Terebratulina gracilis* occurs in the Upper Maastrichtian samples NK4, NK5, NK6, NK7, NK8, and NK9. The species is mainly represented by few individuals, except for sample NK4 where it is among the dominating species. At the Maastrichtian–Danian boundary, the species becomes extinct.

Terebratulina longicollis Steinich 1965
Pl. 7:4–6; Fig. 15A–F

Synonymy. – □ 1909 *Terebratulina locellus* Roemer – Nielsen, p. 160, no. 21. □ 1965 *Terebratulina longicollis* sp.n. – Steinich, pp. 66–72. Figs. 62–75. Pl. 7:2a–d. □ 1972 *Terebratulina longicollis* Steinich – Surlyk, p. 18, Figs. 5, 12–15, 17, 18, Pl. 2:a, b. □ 1979 *Terebratulina longicollis* Steinich – Bitner & Pisera, pp. 75–76, Pl. 3:9–11. □ 1982 *Terebratulina longicollis* Steinich – Surlyk, Fig. 1, Pl. 1e. □ 1984 *Terebratulina longicollis* Steinich – Surlyk & Johansen, Fig. 1.

Material. – Thirty-five complete shells, 41 dorsal and 31 ventral valves. Largest specimen is from the Upper Maastrichtian sample NK5 and has following dimensions: length 3.36 mm; dorsal valve length 2.76 mm; width 2.08 mm; thickness 1.24 mm; foramen width 0.28 mm; and number of ribs 11.

Description. – The shell is small, the outline is pointed oval to elongated subpentagonal, and the maximum width lies near its mid-length. The auricles are well defined and oblique. The shell is biconvex, and the anterior commissure is rectimarginate, and in some large specimens almost uniplicate. The shell surface possesses 10–12 strong, singular, straight, heavily knobbed radial ribs; no ribs are formed by intercalation. The rib sculpture is strongest on the ventral valve. Only a few distinct growth lines are seen. The hinge line is short and the beak is short and suberect. The area is small and distinctly delimited. The foramen is subtriangular and hypothyridid in juveniles. In larger forms, the foramen is large, mesothyridid, circular and limited anteriorly by converging well defined deltidial plates. The pedicle collar is well developed. The inner socket ridges are high, long and thin, while the outer socket ridges are low. The teeth are acute and rather strong. The development of the brachidium is consistent with the development described by Steinich (1965, pp. 68, 71). The brachidium is reminiscent of that of *T. chrysalis*, but in *T. longicollis* the dorsal part of the closed

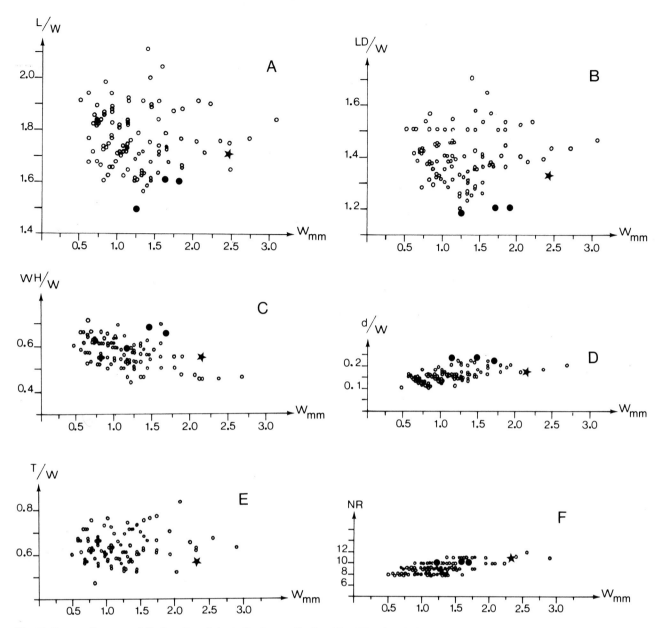

Fig. 15. Scatter diagrams of *Terebratulina* cf. *longicollis*, Lower Danian, Nye Kløv (●; ★: largest specimen, sample NK28) and *T. longicollis* Steinich, Lower Maastrichtian, Rügen (○; data from Steinich 1965, Figs. 62, 63, 65, 68, 69, 71). ☐A. Ratio shell length L (mm) to width W (mm). ☐B. Ratio dorsal valve length LD (mm) to width W (mm). ☐C. Ratio hinge line width WH (mm) to width W (mm). ☐D. Ratio foramen width d (mm) to width W (mm). ☐E. Ratio thickness T (mm) to width W (mm). ☐F. Number of ribs NR.

brachial ring characteristically possesses an acute posteriorly directed nose. The spicular skeleton is heavy. The lophophore was a plectolophe. The inner shell surface is a weak negative relief of the outer surface, and the shell is rather thick.

Remarks. – Large forms of *Terebratulina longicollis* are distinguished from *T. chrysalis* (Schlottheim) in possessing singular ribs and an elongated shell. *T. longicollis* possesses half ring-formed knobs across the ribs, while *T. chrysalis* has tubercles on the ribs. *T. faujasii* (Roemer) differs from *T. longicollis* in possessing a broad subtriangular outline, a rib sculpture consisting of very strong half rings, and a different brachidium.

Specimens from Nye Kløv are, in general, smaller than the largest specimens from the type locality described by

Steinich (1965) from the Upper Maastrichtian of Rügen. The largest Danish specimen is clearly smaller than most specimens from Rügen. However, no differences in morphology and ontogenetic development are to be seen (Fig. 15A–F).

Occurrence. – *Terebratulina longicollis* is found in the Upper Maastrichtian samples NK2, NK3, NK4, NK5, NK6, NK7, and NK8 in a constantly high number of individuals. In the samples NK5, NK6, NK7 and NK8 the species is dominant. *T. longicollis* is an index fossil for the uppermost Campanian *longicollis–jasmundi* Zone (Surlyk 1982, 1984) and is also characteristic for the Maastrichtian. The species is found in the Upper Maastrichtian but it is uncertain whether the species dies out at the Maastrichtian–Danian boundary. In Lower Danian, a new form appears which shows very close affinity to *T. longicollis*. The Lower Danian form of this species is described below as *T.* cf. *longicollis*.

Terebratulina cf. *longicollis* Steinich 1965
Pl. 4:6, 7; Fig. 15A–F

Synonymy. – □ 1984 *Terebratulina* aff. *longicollis* Steinich – Surlyk & Johansen, Fig. 1.

Material. – Four complete shells, two dorsal and one ventral valve. The largest specimen is from the Lower Danian sample NK28 and has the following dimensions: length 3.80 mm; dorsal valve length 3.20 mm; width 1.40 mm; foramen width 0.44 mm; number of ribs 11; and hinge line width 1.32 mm.

Description. – The shell is small, rather thick, and the outline is pointed oval to elongated subpentagonal. The maximum width lies at the mid-length of the shell, and the auricles are well defined. The shell is biconvex, and the shell surface shows 9–11 single straight, strong, heavily sculptured, radial ribs. There is no intercalation of the ribs. The rib sculpture consists of narrow transverse half rings, the ribs are relatively narrow, and the interspaces are as wide as the ribs. The beak is suberect, high and blunt, and the pedicle collar is well developed. The foramen is large, subtriangular, and mesothyridid; the deltidial plates and the area are well defined. The inner socket ridges are high, thin and long, and the teeth are acute and strong. The cardinal process is a transverse plate, and the crura are relatively short, strong, and converging midventrally. The crural processes unite mid-dorsally in a rather wide bridge. The descending branches and the crural processes form together a ring-like loop, as is characteristic for the genus *Terebratulina*, and the dorsal part of the brachial ring possesses a characteristic posteriorly directed process. The spicular skeleton is strong and the anterior part of the skeleton lies as a collar close to the anterior margin of the dorsal part of the brachidium. The lophophore was probably a plectolophe. The shell inner surface is negative relief of outer.

Remarks. – *Terebratulina* cf. *longicollis* from the Lower Danian is very similar to the Maastrichtian form of *T. longicollis* Steinich as described by Steinich (1965) and Surlyk (1969). In their internal morphology the specimens from Nye Kløv are similar to *T. longicollis*. They have, however, slightly lower ratios of length to width (L/W) than the specimens from Rügen, and the ribs are not as wide as in the original diagnosis for *T. longicollis*. The species from the Lower Danian may be a new species, but due to the sparse material present nothing definite can be stated, and the material is here referred to as *T.* cf. *longicollis*. The material from Nye Kløv is for comparison shown in scatter diagrams for *T. longicollis* (Fig. 15A–F).

Occurrence. – *Terebratulina* cf. *longicollis* occurs only in a few Lower Danian samples (NK27, NK28, NK30) and is represented by only a few individuals. It has its first appearance about 8 m above the Maastrichtian–Danian boundary.

Terebratulina kloevensis n.sp.
Pls. 6:1–5; 8:1, 4, 5; Fig. 16A–H

Synonymy. – □ 1984 *Terebratulina* n.sp. – Surlyk & Johansen, Fig. 1.
Derivation of name. – After the name of the type locality, Nye Kløv.

Holotype. – MGUH 16924, Pl. 6:1A, B; Lower Danian, Nye Kløv, sample NK30. Type locality: Nye Kløv, north of Thisted, northwestern Denmark. Type horizon: Lower Danian.

Material. – Thirty-three complete shells, 33 dorsal and 33 ventral valves and a number of fragments. The largest specimen is from the Lower Danian sample NK29 and has the following dimensions: length 3.60 mm; dorsal valve length 2.88 mm; width 2.80 mm; thickness 1.40 mm; foramen width 0.44 mm; number of ribs 28. Measurements of the holotype: length 2.64 mm; dorsal valve length 2.12 mm; width 2.32 mm; thickness 0.92 mm; foramen width 0.30 mm; number of ribs 23.

Diagnosis. – Small shell with elongate subpentagonal outline. Shell surface with 20–30 finely granulated ribs. Rib sculpture and pattern differ from ventral to dorsal valve. At ventral valve ribs are narrow, covered with small, well defined granules. Ribs run here along the valve length and are in profile slender acute ridges. Ribs of dorsal valve are wide, low and spread fan-like, and granulation is very faint. Foramen relatively large, submesothyridid and brachidium characteristically ring-shaped.

Description. – The shell is small, with an outline changing during ontogeny from pointed subtriangular through rounded subtriangular to elongated subpentagonal. Throughout ontogeny, maximum width is anterior to the mid-length of the shell. The auricles are small. The ratios L/W, LD/W, T/W and d/W are shown in Fig. 16A–D. The adult shell is ventri- biconvex while juveniles are almost plano-convex. The anterior commissure is rectimarginate. The shell surface displays numerous ribs. The dorsal and ventral valves have slightly different rib patterns and sculptures. At a shell width of 0.80 mm, 6–7 ribs are present, and new ribs are rapidly added by intercalation. The large specimens contain at least 28 ribs (Fig. 16E, F). The rib sculpture consists of granules arranged across the ribs and are most distinctly seen on the ventral valve. The ventral valve ribs are narrow straight, slender, and form in profile acute ridges. They become slightly wider during growth. The dorsal valve ribs are straight, low and wide, and arranged in a fan-like pattern. The hinge line is relatively short and straight, the beak is suberect to erect, and the area is narrow. The foramen is relatively large, subtriangular and hypothyridid in the juvenile stages, and oval and submesothyridid in the later stages. Juvenile deltidial plates are relatively high, narrow ridges that are posteriorly disjunct. Later in ontogeny, the foramen is limited anteriorly and laterally by discrete triangular plates converging dorsally. Posteriorly, the foramen is confined by a well-developed pedicle collar. The beak is often attrite, and the hinge is relatively strong. The teeth are

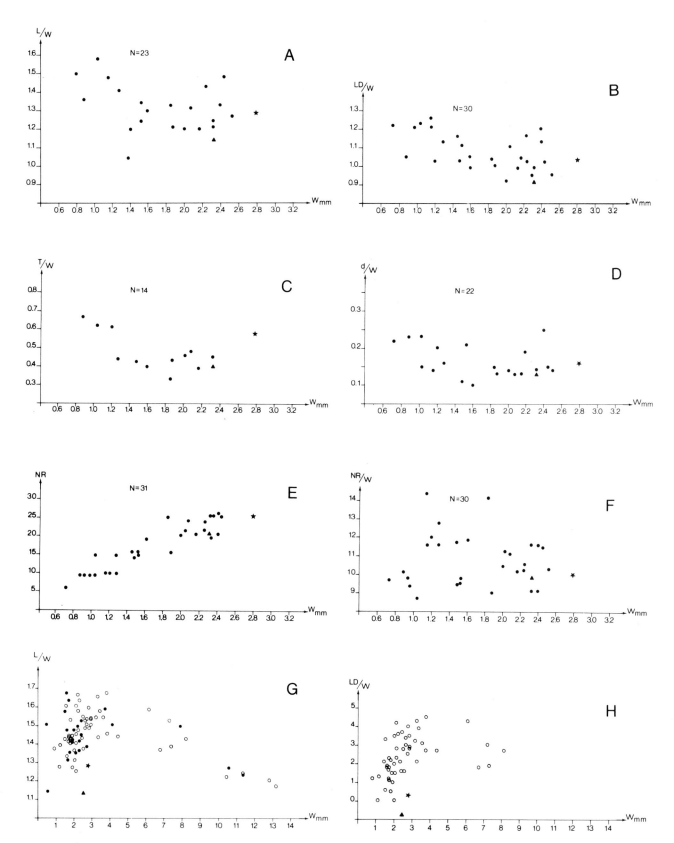

Fig. 16. Scatter diagrams of *Terebratulina kloevensis* n.sp., Lower Danian, Nye Kløv (▲: holotype, sample NK30, MGUH 16924; ★: largest specimen, sample NK29), and (in G and H) *T. chrysalis* (Schlottheim), Lower Maastrichtian, Rügen (○, ●; data from Steinich 1969, Figs. 44, 45). ☐A. Ratio shell length L (mm) to width W (mm). ☐B. Ratio dorsal valve length LD (mm) to width W (mm). ☐C. Ratio thickness T (mm) to width W (mm). ☐D. Ratio foramen width d (mm) to width W (mm). ☐E. Number of ribs NR. ☐F. Ratio number of ribs to width NR/W. ☐G. ☐G. Ratio shell length L (mm) to width W (mm). ☐H. Ratio dorsal valve length LD (mm) to width W (mm).

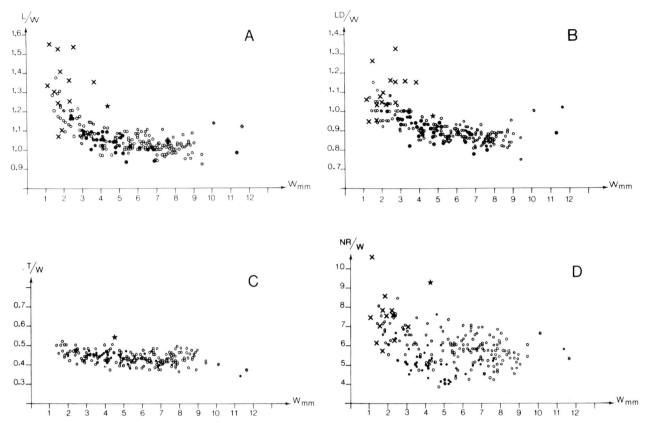

Fig. 17. Scatter diagrams of *Terebratulina* aff. *rigida* from the Lower Danian, Nye Kløv (×; ★: largest specimen, sample NK29), and *T. gracilis* (Schlottheim), Lower Maastrichtian, Rügen (○, ●; ; data from Steinich 1965, Fig. 95, 96, 98, 100). □ A. Ratio shell length L (mm) to width W (mm). □ B. Ratio dorsal valve length LD (mm) to width. □ C. Ratio thickness T (mm) to width W (mm). □ D. Ratio number of rib NR to width W (mm).

medium-sized and blunt. The inner socket ridges are thin, high and converging posteriorly. They reach well behind the posterior margin of the dorsal valve. The cardinal process is medium-sized and flat. The brachidium is similar to that of other *Terebratulina* species. At a shell width of about 1.00 mm, two strong, midventrally converging crura are developed anterior to the inner socket ridges. The material does not allow any further study of the ontogenetic development of the brachidium, but at a shell width of 2.80 mm, relatively long, thin, midventrally converging crura were observed. At this stage, the crural pocesses are fused in a wide, broad ventral arch, and the descending branches are fused in a narrow, acute dorsal arch. Together, the crural processes and the descending branches form a subtriangular ring (Pl. 6:2). The lophophore was probably a plectolophe. The inner surface of the shell reflects a negative relief of the outer surface. The shell is relatively thick.

Remarks. – Juveniles of *Terebratulina chrysalis* (Schlottheim) differ from *Terebratulina kloevensis* n. sp. of the same size in having strongly developed shoulders and in having straight, strongly granulated ribs on both the ventral and dorsal valves. Furthermore, L/W and LB/W are generally lower for *T. kloevensis* than for *T. chrysalis* as shown in Fig. 16G, H. *T. subtilis* Steinich differs from *T. kloevensis* in having a much larger number of ribs. At a shell width of 2.00 mm, *T. subtilis* has about 50 ribs, whereas *T. kloevensis* has 25 ribs. Furthermore, *T. subtilis* is characterised by a rib sculpture, consisting of numerous small grains on top of the low, wide ribs.

Occurrence. – *Terebratulina kloevensis* occurs in the Lower Danian samples NK27, NK29, and NK30. The species is confined to the Lower Danian, and is here limited to the upper part of the section. *T. kloevensis* occurs only in a few samples, but is then abundant.

Terebratulina aff. *rigida* (Sowerby 1821)
Pls. 5:1–4; 8:2; Figs. 14A–D, 17A–F

Synonymy. – □ 1821 *Terebratulina rigida* – Sowerby, p. 69, Pl. 536:2. □ 1866 *Terebratulina rigida* Sow.sp. – Schloenbach, p. 287, Pl. 37:10–17. □ 1984 *Terebratulina* aff. *rigida* (Sowerby) – Surlyk & Johansen, Fig. 1.

Material. – Twenty-eight complete shells, 15 dorsal and 13 ventral valves. The largest specimen is from the Lower Danian sample NK29 and has the following dimensions: length 5.32 mm; dorsal valve length 4.32 mm; width 4.40 mm; thickness 2.40 mm; foramen width 0.56 mm; number of ribs 40; and hinge line width 1.60 mm.

Description. – The shell is relatively large compared to other cancellothyridid species, and the outline is subpentagonal to subtriangular. The maximum width lies in juveniles around the mid-length of the shell and moves, during ontogeny, to a more anterior position. The auricles are small but well defined and the shell is slightly ventri-biconvex. The anterior commissure is straight to uniplicate. At a shell width of 2.00 mm, about 10 ribs are present, and at least 40 ribs are present at 4.40 mm. New ribs are formed by intercalation. The ribs are straight and coarsely knobbed with slight backward reflexion in large specimens. The rib sculpture consists of prominent high, transverse half-rings arranged in closely spaced rows along the ribs. The ribs are always wider than the interspace between them, and the width of the ribs increases during ontogeny. Numerous distinct growth lines are present. The hinge line is short and laterally rather oblique. The beak is relatively short, erect to slightly incurved and blunt. The area is very narrow and triangular in juveniles, and in large specimens a broad palintrope limits the area to a very narrow ridge. The foramen is rather large, low, broad, subtriangular and hypothyridid in juveniles; during ontogeny, the outline of the foramen changes from elongated oval to oval, and the foramen changes from submesothyridid to mesothyridid. The deltidial plates are high and triangular, limiting the foramen laterally and anteriorly, but they are not fused anteriorly in any of the observed specimens. The hinge and the brachidium are strong, and the inner socket ridges are short, high and mutually convex with a long ventral face. The inner sockets are very deep, and the outer socket ridges are short, thin and relatively low. The cardinal process is slightly concave, and a platform is inserted between the inner socket ridges. The crura develop in front of the anterior margin of the inner socket ridges, and the descending branches are short, strong and rapidly converging in a midventral direction. Anterior to the midline of the shell a rounded subtriangular brachial ring is formed. The descending branches unite mid-dorsally under a wide angle into a relatively wide transverse band. The anterior part of the junction is prolonged into a characteristically broad, subtriangular, ventrally directed lobe. The crural processes unite ventrally into a wide, slightly concave bridge. The spicular skeleton was heavy and the lophophore was a plectolophe. The spicular skeleton in the anterior body chamber was wide and blanket-like and placed as a broad collar around the anterior edge of the dorsal part of the brachidium. The arms of the spicular skeleton formed a plectolophe with wide brachial lips and short filaments (Pl. 5:1C, C, E). The inner surface of the shell is smooth, and the anterior margins are negative relief of outer surface. The shell is rather thick.

Remarks. – *Terebratulina gracilis* (Schlottheim) and *Terebratulina rigida* have often been confused in earlier literature. *T. gracilis* has been identified, quoted and described by many (see Steinich 1965, list of synonyms for *T. gracilis*), but the question is how many of the quoted specimens ought to be referred to as *T. rigida* instead. Schloenbach (1866) distinguished between *T. rigida* and *T. gracilis* and mentions the problems of confusing the two species (Schloenbach 1866, pp. 17–22). The type specimen of *T. gracilis* described by Sowerby (1821) and illustrated by, among others, Schloenbach (1866, Fig. 18a, b), possesses a concavo-convex shell and an incurved beak, but the ribs appear less reflexed than in other *T. gracilis* specimens described (e.g. Steinich 1965, Pl. 11:1a–d). Schloenbach (1866) combines four different *T. gracilis*-like forms from Upper Cretaceous into *T. rigida*, and Steinich (1965) mentions that these forms may be synonyms of *T. gracilis* (Schloenbach 1866, Pl. 38:10–17). The material from the Lower Danian, Nye Kløv, is, however, to be distin-

guished from *T. gracilis* and has close affinities to the original *T. rigida* (Schloenbach 1866). For present purposes it is preferred to retain *T. rigida* as a separate species. As the interior morphology of the holotype of *T. rigida* is unknown, the Lower Danian specimens are referred to as *T. aff. rigida*. *T. gracilis* from Nye Kløv is different from *T. aff. rigida* in having a subcircular outline, a planoconvex to concavo-convex shell, a very acute and incurved beak and in having a constant foramen width of 0.2–0.3 mm (foramen width of *T. aff. rigida* varies between 0.06 and 0.56 mm), and in possessing distinctly reflexed ribs.

T. chrysalis (Schlottheim) differs from *T. aff. rigida* in the more elongate outline, the larger rib number and in having ribs that do not widen during growth. The largest specimen of *T. aff. rigida* is shown is scatter diagrams for *T. chrysalis* for comparison (Fig. 14A–D) and L/W, LD/W, T/W and NR/W for *T. aff. rigida* is shown in scatter diagrams for *T. gracilis* for comparison (Fig. 17A–F).

Occurrence. – *Terebratulina* aff. *rigida* occurs in the Lower Danian samples NK23, N24, NK26, NK27, and NK29. The species does not occur in the Upper Maastrichtian of Denmark. This is the first recorded occurrence of the species in the Lower Danian.

Subfamily Chlidonophorinae Muir-Wood 1959

Genus *Gisilina* Steinich 1963

Type species. – *Terebratula gisii* Roemer 1840, by original designation.

Gisilina jasmundi Steinich 1965
Pl. 8:3

Synonymy. – □ 1965 *Gisilina jasmundi* sp.n. – Steinich, pp. 110–115. Figs. 149–161, Pl. 16:1a–d, 2. □ 1972 *Gisilina jasmundi* Steinich – Surlyk, p. 18, Figs. 2, 5, 13, 15a, 16, 18. □ 1982 *Gisilina jasmundi* Steinich – Surlyk, Fig. 1, Pl. 2:a,b,c. □ 1984 *Gisilina jasmundi* Steinich – Surlyk & Johansen, Fig. 1.

Material. – The material from Nye Kløv contains a large number of unidentified *Terebratulina* and probably also *Gisilina* fragments, but no complete large shells. Only a few valves can with certainty be referred to *Gisilina jasmundi*. Fragments with the same rib pattern and sculpture as *G. jasmundi* occur, but because they closely resemble *T. chrysalis*, it has not been possible to identify them with confidence.

A dorsal valve that may certainly be referred to *G. jasmundi* is from the Upper Maastrichtian sample NK5 and has the following dimensions: length 3.24 mm; width 2.80 mm; foramen width 0.68 mm; number of ribs 16.

Description. – *Gisilina jasmundi* has been thoroughly described by Steinich (1965). Characteristically it possesses 12–18 straight, coarsely knobbed singular radial ribs and a relatively flat shell. The rib sculpture varies between coarse knobs on top of the ribs and prominent transverse half-rings arranged across the ribs.

The present specimen shows that the shell outline is subpentangular and the maximum width is posterior to the midline of shell. The auricles are small but distinct, and the hinge line is relatively short and almost straight. The rib surface possesses 16 straight, single and strongly sculptured wide ribs. The ribs in the late growth stages are much wider than in young stages.

The shell is relatively thick.

Remarks. – Small specimens of *Gisilina jasmundi* are difficult to distinguish from small specimens of *Gisilina gisii* (Roemer) *Gisilina gisii* differs, however, in its smooth ribs and more biconvex shell.

Occurrence. – *Gisilina jasmundi* occurs in the Upper Maastrichtian, but with certainty only in sample NK5.

The species is index fossil for the lower Lower Maastrichtian *jasmundi–acutirostris* Zone of Surlyk (1982, 1984). *G. jasmundi* is furthermore characteristic throughout the Lower Maastrichtian. The species becomes extinct at the Maastrichtian–Danian boundary, and does not occur in the Lower Danian.

Genus *Rugia* Steinich 1963

Type species. – *Rugia tenuicostata* Steinich 1963, by original designation.

Rugia acutirostris Steinich 1965
Pl. 7:1A, B

Synonymy. – □ 1965 *Rugia acutirostris* sp.n. – Steinich, pp. 122–124, Figs. 175–178, Pl. 14:1a–d. □ 1972 *Rugia acutirostris* Steinich – Surlyk, p. 18, Figs. 2, 5, 14, 18. □ 1979 *Rugia acutirostris* Steinich – Bitner & Pisera, pp. 77–78, Pl. 3:1–3. □ 1982 *Rugia acutirostris* Steinich – Surlyk, Fig. 1, Pl. 1:i, j. □ 1984 *Rugia acutirostris* Steinich – Surlyk & Johansen, Fig. 1.

Material. – One complete shell, two dorsal and two ventral valves. The largest specimen is represented by a broken dorsal valve from the Upper Maastrichtian sample NK4 and has the following dimensions: length 2.2 mm; width 1.8 mm; and number of ribs >30.

Description. – The material is referred to genus *Rugia* Steinich due to the hypothyridid foramen, the short, straight hinge line, and the long, tapering deltidial plates. The specimens from Nye Kløv have a very pointed oval outline, a very acute beak, a very small circular foramen and 20–40 finely knobbed ribs. The rib sculpture consists of rather prominent granules, new ribs are formed by intercalation, and the ribs are slightly reflexed. These characters are consistent with those described by Steinich (1965) and Surlyk (1972) for *Rugia acutirostris*.

Remarks. – *Rugia acutirostris* differs from other *Rugia* species in its very acute beak, very small circular foramen, and large number of ribs.

Occurrence. – The species is represented by a low number of individuals in the Upper Maastrichtian samples NK4, NK7 and NK8.

Rugia acutirostris is an index fossil for the lower Lower Maastrichtian *acutirostis–spinosa* Zone of Surlyk (1982, 1984), but occurs also in the remaining part of the Maastrichtian. The species becomes extinct at the Maastrichtian–Danian boundary.

Rugia tenuicostata Steinich 1963
Pl. 7:2; Fig. 18F, G

Synonymy. – □ 1963b *Rugia tenuicostata* sp.n. – Steinich, pp. 737–739, Figs. 6–8. □ 1965 *Rugia tenuicostata* Steinich – Steinich, pp. 116–121, Figs. 162–174, Pl. 11:3a–d, 4. □ 1972 *Rugia tenuicostata* Steinich – Surlyk, pp. 18–19, Figs. 2, 5, 12, 16. □ 1979 *Rugia tenuicostata* Steinich – Bitner & Pisera, p. 77, Pl. 4:3–6, Pl. 5:3. □ 1982 *Rugia tenuicostata* – Surlyk, Fig. 1, Pl. 1f,g,h. □ 1984 *Rugia* cf. *tenuicostata* Steinich – Surlyk & Johansen, Fig. 1.

Material. – One complete shell, one dorsal and two ventral valves. The largest specimen is represented by a dorsal valve from the Upper Maastrichtian sample NK4 and has the following dimensions: length 2.12 mm; width 1.92 mm and the number of ribs 28.

Description. – The species is referred to the genus *Rugia* Steinich due to the hypothyridid foramen, the short and pointed beak, the short, straight hinge and the long, tapering deltidial plates. The material is consistent with the *R. tenuicostata* described by Steinich (1963, 1965) and Surlyk (1972). The specimens from Nye Kløv have the same characteristic rib sculpture consisting of small triangular tubercles aligned on top of the ribs, which gives the profile of the ribs a fretsaw-like appearance.

Remarks. – *Rugia acutirostris* Steinich is distinguished from *R. tenuicostata* in possessing a much more acute beak, and a larger number of finely knobbed ribs.

Rugia flabella n.sp. is distinguished from *R. tenuicostata* in having finely knobbed ribs that are more reflexed and in having a lower length-to-width ratio (LD/W). *Rugia latronis* n.sp. differs from *R. tenuicostata* in a lower rib number and in having crura which are not fused with the inner socket ridges.

In connection with the description of *R. flabella* elsewhere in this paper, scatter diagrams of L/W and LD/W for *R. tenuicostata* from the Lower Maastrichtian of Rügen are compared with those of *R. flabella* and *R. latronis* (Fig. 18F, G).

Occurrence. – The species occurs in the Upper Maastrichtian samples NK4 and NK6 in a low number of individuals.

Rugia tenuicostata is index fossil for the Late–latest Campanian *tenuicostata–longicollis* Zone of Surlyk (1983, 1984), and for the latest Early Maastrichtian *tenuicostata–semiglobularis* Zone. This is the first record of the species from the latest Maastrichtian. The species dies out at the Maastrichtian–Danian boundary.

Rugia flabella n.sp.
Pl. 9:1–7; Fig. 18A–G

Synonymy. – □ 1984 *Rugia* n.sp. 1 – Surlyk & Johansen, Fig. 1.

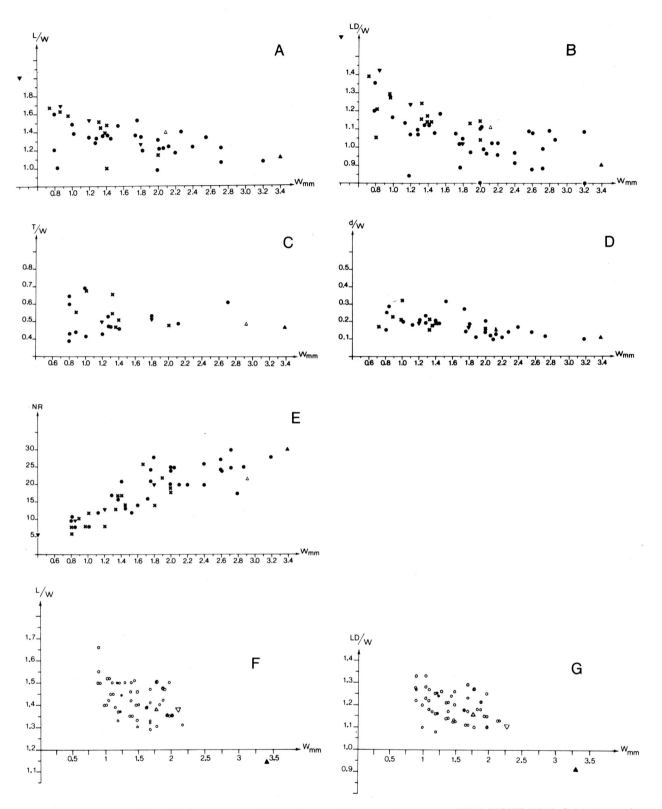

Fig. 18. Scatter diagrams of *Rugia flabella* n.sp. (● in A–F; ▲: holotype and largest specimen, sample NK19, MGUH 16941), *R. latronis* n.sp. (×; △: holotype and largest specimen, sample NK26, MGUH 16948), and *R.* sp. (▼; sample NK24), Lower Danian, Nye Kløv, and (in G–H) *R. tenuicostata* Steinich, Maastrichtian, Rügen (○, ●; data from Steinich 1965, Figs. 162, 163). □A. Ratio shell length L (mm) to width W (mm). □B. Ratio dorsal valve length LD (mm) to width W (mm). □C. Ratio thickness T (mm) to width W (mm). □D. Ratio foramen width d (mm) to width W (mm). □E. Number of ribs NR. □F. Ratio dorsal valve length LD (mm) to width W (mm). □G. Ratio dorsal valve length LD (mm) to width W (mm).

Derivation of name. – Latin *flabella*, a little fan; referring to the growth pattern of the ribs.

Holotype. – MGUH 16941, Pl. 9:1A, B; Lower Danian, Nye Kløv, Sample NK19. Type locality: Nye Kløv, north of Thisted, northwestern Denmark. Type horizon: Lower Danian.

Material. – The material consists of 46 complete shells, 83 dorsal and 43 ventral valves and a number of fragments. The holotype is the largest specimen present, and has the following dimensions: length 3.88 mm; dorsal valve length 3.12 mm; width 3.40 mm; thickness 1.60 mm; foramen width 0.44 mm; number of ribs 30.

Diagnosis. – Shell large, biconvex and relatively thick, with broad subtriangular outline. Beak erect to slightly incurved and anterior commissure rectimarginate to incipiently sulcate. Shell surface possesses 20–30 low, finely knobbed, gently backward-bending ribs. Foramen relatively large, limited laterally by two deltidial plates formed as irregular ridges. Teeth relatively large and blunt with grooved dorsal face, cardinal process large and bulbous. Descending branches converge dorsally and fuse mid-dorsally, forming an acute angle.

Description. – *Rugia flabella* is comparatively large for the genus (cf. Steinich 1965, p. 115). The outline is in the juveniles elongated oval to subpentangular, later pointed oval, and in the largest specimens pointed broad subtriangular. The maximum width is anterior to mid-length of the shell throughout ontogeny. The ratios L/W, LD/W, T/W and d/W are shown in Fig. 18A–D. The auricles are small and poorly defined, and the shell is slightly ventri-biconvex Furthermore, the juveniles are the most biconvex. The anterior commissure is rectimarginate to incipiently sulcate. At a shell width of about 0.50 mm, 5–6 ribs occur, and the number of ribs increases rapidly by intercalation. Large shells contain at least 30 ribs at the shell margin (Fig. 18E). Laterally, the ribs are curved gently backwards, giving the shell surface a fan-like appearance. The ribs maintain their width during ontogeny. The ribs are low and sculptured by crowded, narrow half-rings giving the rib surface a finely knobbed appearance. The interspace between the ribs is almost as wide as the ribs. Five to six distinct growth lines can be seen, and these often give the shell surface a flaky appearance. The beak is often attrite, short, suberect to erect, and may be slightly incurved in large forms. The area is small but often recessed. The foramen is hypothyridid and in the juveniles oval to subcircular. During ontogeny the foramen becomes subtriangular. The foramen is confined laterally by two narrow deltidial plates formed as irregular ridges which do not restrict the foramen anteriorly. The pedicle collar is well developed and comprises about half the length of the beak. A deep pedicle groove is present. The hinge is well developed. The teeth are relatively large, blunt, triangular and dorsally converging. The dorsally directed face of the teeth contains a well developed groove. The inner socket ridges are high, outwardly bending plates, and the sockets are equally deep. The outer socket ridges are fused with the crural bases. The cardinalia reach behind the

posterior margin of the dorsal valve. The cardinal process is relatively large and bulbous. The crura are fused with the inner socket ridges and continue ventrally into long, slender and laterally compressed descending arms. The descending branches converge in a ventral direction at a high angle to the bases of the arms. At a shell width of about 1.50 mm, the lophophore was a schizolophe, and at 2.70 mm a zygolophe. The largest forms were possibly plectolophous. The spicular skeleton is well developed, and the cirri were long. The inner surface of the shell shows at the margin a negative relief of the ribs. The remainder of the shell interior is either smooth or shows up to 10 shallow furrows that run the length of the valve. A low median ridge is often present in the dorsal valve. The shell is relatively thick.

Remarks. – The species undoubtedly belongs to the genus *Rugia* Steinich (Steinich 1963b, pp. 735–738), because of the form of the brachidium, the hypothyridid foramen and the irregular deltidal plates. *Rugia tenuicostata* Steinich can be distinguished from *Rugia flabella* in having straighter ribs, being markedly longer than wide, having collar-like deltidial plates, and having a less bulbous cardinal process. *Rugia acutirostris* Steinich is distinguished from *R. flabella* in having a much smaller foramen width and a more acute beak. *Rugia tegulata* Surlyk can be distinguished from *R. flabella* in having a rib sculpture consisting of plate-like scales.
  Rugia latronis n.sp. differs from *R. flabella* in having coarser-sculptured and more irregular ribs, and in being more elongate subtriangular in outline.
  L/W and LD/W for the holotype of *Rugia flabella* and the holotype of *Rugia latronis* n.sp. are for comparison plotted in scatter diagrams for *Rugia tenuicostata*. While *R. latronis* falls within the range of variation of *R. tenuicostata*, *R. flabella* does not (Fig. 18F, G).

Occurrence. – The species occurs in a relatively large number of Lower Danian samples (NK17, NK18, NK19, NK20, NK21, NK23, NK24, NK26, NK27, NK28, NK30) and in very variable numbers. It may be represented by a few specimens, but more commonly it dominates the brachiopod assemblage, as in samples NK19, NK21, NK24, and NK28. *R. flabella* is one of the first brachiopod species to appear in the Lower Danian at Nye Kløv.

Rugia latronis n.sp.
Pl. 10:1–3; Fig. 18A–G

Synonymy. – □ 1984 *Rugia* n.sp. 2 – Surlyk & Johansen, Fig. 1.

Derivation of name. – Latin *latronis*, a thief. The species has the appearance of having stolen characters randomly from other *Rugia* species.

Holotype. – MGUM 16948, Pl. 10:1A, B; Lower Danian, Nye Kløv, Sample NK26. Type locality: Nye Kløv, north of Thisted, northwestern Denmark. Type horizon: Lower Danian.

Material. – Nineteen complete shells, seven dorsal and three ventral valves and a number of fragments. The largest specimen is the holotype and has the following dimensions: length

2.92 mm; dorsal valve length 2.36 mm; width 2.12 mm; thickness 1.04 mm; foramen width 0.32 mm, and number of ribs 21.

Diagnosis. – Shell small, with elongated subtriangular outline. Surface contains 20–22 irregular intercalated, rather strongly sculptured ribs with sculpture of prominent half-rings. Characteristic alternation between strong and weak ribs. Foramen relatively large, hypothyridid; deltidial plates low, irregular ridges; area narrow triangular. Cardinal process flat. Long, slender crura not fused with inner socket ridges. Crura converge antero-ventrally, and descending branches dorsally.

Description. – The shell is small with an elongated subpentangular outline in the juvenile stages, becoming pointed oval to elongated subtriangular in larger forms. The maximum width is close to the anterior shell margin, and the auricles are small. The anterior commissure is rectimarginate to broadly sulcate. The adult shells possess 6–7 distinct growth lines. At a shell width of about 0.80 mm, 5–6 straight radial ribs are formed, the number of which increases rapidly by intercalation. At shell widths of about 2.10 mm, 20–22 ribs are present. The sculpture of the ribs is rather strong, consisting of prominent half-rings, and where the ribs cross the growth lines, the half-rings appear almost scale-like. There is a characteristic alternation between strong and weak ribs, and the ribs are clearly wider than the interspace between them. The growth pattern of the ribs is rather irregular and gives them a discontinuous appearance. The ribs protrude well beyond the anterior shell margin. The hinge line is short and straight, and the beak is short, erect and relatively acute. The area is narrow and triangular, and the foramen is hypothyridid and relatively large. The outline of the foramen is semi-circular in juveniles but becomes triangular during ontogeny. The deltidial plates are developed as two low irregular ridges. A pedicle collar is well developed and is transsected by a shallow pedicle groove. The hinge is relatively strong. The inner socket ridges are short but strong, and the cardinal process is flat. The crura are long and slender and are not fused with inner socket ridges in juvenile and medium growth stages. It has not been possible to investigate the brachidial development in further detail, but through a hole in the ventral valve of the holotype it was possible to observe two long, thin crura rapidly converging in an antero-ventral direction. The distal ends of the crura are flattened in a dorso-ventral direction with the dorsally directed part converging. The material does not allow for a definite statement on the adult lophophore, but it was probably a subplectolophe. The inner surface of the shell is a negative relief of the outer surface.

Remarks. – *Rugia tenuicostata* Steinich differs from *Rugia latronis* n.sp. in having straight ribs and a larger number of them (*R. tenuicostata* has 32 ribs at a shell length of 2.80 mm, while *R. latronis* has 22 at 2.90 mm), and in the collar-like deltidial plates. Furthermore, in *R. tenuicostata* the crura are fused with the inner socket ridges. L/W and LD/W for *R. latronis* are for comparison plotted in the scatter diagrams for *R. tenuicostata* (Fig. 18F, G). *R. acutirostris* Steinich differs from *R. latronis* in having a smaller foramen, a larger number of

ribs, and a much more acute beak. *R. tegulata* Surlyk differs from *R. latronis* in having a rib sculpture consisting of plate-like scales.

R. flabella sp.n. can be distinguished from *R. latronis* in its relatively larger size and in having a more regular pattern of gently backward-bending, finely knobbed ribs, and a broader triangular outline of the shell. The scatter plots for the ratios L/W, LD/W, T/W, d/W and NR/W for *R. latronis* are for comparison shown in scatter diagrams for *R. flabella* (Fig. 18A–E).

Occurrence. – *Rugia latronis* n.sp. is found in the Lower Danian samples NK23, NK24, NK25, and NK26, and has its first apperance at the Nye Kløv locality around 6.00 m above the Maastrichtian–Danian boundary.

R. latronis is represented only by a small number of individuals in the samples in which it occurs. This is the first recorded occurrence of the species in the Lower Danian.

Rugia sp.
Pl. 10:4, 5; Fig. 18A–E

Synonymy. – ☐ 1984 *Rugia*(?) sp. – Surlyk & Johansen, Fig. 1.

Material and occurrence. – The material, which consists of two complete shells, one dorsal and three ventral valves, was found in only one sample from the Lower Danian at Nye Kløv, sample NK24. The dimensions of the largest specimen present are: length 2.32 mm; dorsal valve length 1.84 mm; width 1.80 mm; thickness 0.92 mm; foramen width 0.92 mm, and number of ribs 22.

Description. – The shell is small and rather thick. The outline is elongated triangular in the juvenile stages and triangular to subpentangular in the adult stages. The maximum width is at the anterior shell margin. The auricles are small and indistinct. The shell is biconvex to almost planoconvex with the ventral valve showing the least convexity. The shell surface possesses 20–22 straight and finely knobbed ribs, and new ribs are formed by intercalation. Only a few distinct growth lines can be observed. Five to six ribs are formed at a shell width of 0.40 mm, and 20–22 ribs at a shell width of 1.80 mm. The hinge line is short and almost straight. The beak is erect in the juvenile stages and erect to slightly incurved in later stages, where it also becomes more acute. The area is relatively broad and triangular, and the deltidial plates are two narrow straight ridges that limit the subtriangular, hypothyridid foramen laterally. A pedicle collar is well developed. A trace of a pedicle groove occurs anteriorly in the pedicle collar. The hinge is well developed. The inner socket ridges are relatively high and the cardinal process is large but flat. The restricted nature of the material precludes detailed survey of the development of the brachidium, but at a shell length of 1.84 mm two slender, ventrally converging crura can be seen. They are laterally compressed and are not fused with the inner socket ridges. The inner surface of the shell is a negative relief of the outer surface.

Remarks. – The material is too rare to allow any confident generic assignment. The long and slender crura, the hypothyridid foramen and the similarity to other *Rugia* species found in the Lower Danian of Nye Kløv suggest, however, a close relationship to the genus *Rugia*. *Rugia* sp. may be a subspecies or a variant of *Rugia flabella* n.sp., but the outline of the latter is more broadly subtriangular, and furthermore the ribs are in this species bending backwards. *R. latronis* n.sp. differs from *Rugia* sp. in possessing more irregular and coarsely sculptured ribs and a weaker hinge. L/W, LD/W, NR/W and T/W ratios for *Rugia* sp. are for comparison plotted in scatter diagrams for *R. flabella* (Fig. 18A–E).

Subfamily Eucalathinae Muir-Wood 1965

Genus *Meonia* Steinich 1963

Type species. – *Terebratulina semiglobularis* Posselt 1894, by original designation.

Meonia semiglobularis (Posselt 1894)
Pl. 2:3, 4, 5

Synonymy. – ☐ 1894 *Terebratulina semiglobularis* n.sp. – Posselt, Pl. 3:10–13. ☐ 1909 *Terebratulina semiglobularis* Posselt – Nielsen, p. 161 no. 24, Pl. 1:35. ☐ 1963b *Meonia semiglobularis* (Posselt) – Steinich, pp. 733–734, Pl. 1–3. ☐ 1965 *Meonia semiglobularis* (Posselt) – Steinich, pp. 46–52, Figs. 35–43, Pl. 7:3a–d, 4, 5. ☐ 1970b *Meonia semiglobularis* (Posselt) – Surlyk, pp. 7–16, Figs. 1–3. ☐ 1972 *Meonia semiglobularis* (Posselt) – Surlyk, p. 26, Figs. 2, 5, 11, 13, 15, 17, Pl. 1:a–f. ☐ 1984 *Meonia semiglobularis* (Posselt) – Surlyk & Johansen, Fig. 1.

Material. – Forty-one complete shells, 168 dorsal and 185 ventral valves. The largest specimen is from the Upper Maastrichtian sample NK6 and has the following dimensions: length 3.00 mm; dorsal valve length 2.76 mm; width 2.32 mm; thickness 1.28 mm; foramen width 0.16 mm; and number of ribs 17.

Description. – The shell is small and thick, and the outline is elongated semicircular. The maximum width lies anterior to the midline of the shell. The auricles are indistinct, and the hinge line is broad and straight. The shell is plano-convex with an almost semiglobular ventral valve, and some large forms are almost concavo-convex. The anterior commissure is rectimarginate to sulcate. The shell surface possesses about 20 strong, radial, smooth and straight singular ribs. The ribs are rounded in cross section in the juvenile forms and later they change to almost triangular. No distinct growth lines are observed. The beak is low, suberect to erect and incurved in large forms, and the area is small and clearly confined. The foramen is hypothyridid in juveniles but changes to mesothyridid during growth. The deltidial plates are small and triangular. The pedicle collar is narrow. The hinge is very strong with well developed inner socket ridges, which are incurved in a median direction. The hinge teeth are flat, blunt and medially directed. The brachidium is well developed and consists of broad, ventrally converging strong crura that continue into slender descending branches which fuse into a horizontal dorsal bridge. The crural processes are short and spoon-shaped. The brachidium is similar to that of *Rugia tenuicostata* but is more strongly built

(Steinich 1965, p. 121, Fig. 174). The spicular skeleton is heavy, and the lophophore was a schizolophe. The inner surface of the shell is smooth.

Occurrence. – *Meonia semiglobularis* is found in samples from the Upper Maastrichtian (NK3, NK4, NK5, NK6, NK7, NK8 and NK9), where it dominates, and also in the 3 cm thick clay bed which marks the Maastrichtian–Danian boundary and forms the basal Danian bed. The latter occurrence may be due to reworking, because the shell fragments present in the clay bed are worn and of a tinted colour.

M. semiglobularis is an index fossil for the lower Upper Maastrichtian *semiglobularis–humboldtii* Zone (Surlyk 1982, 1984) and is also characteristic in the remaining part of the Upper Maastrichtian.

The species dies out at the Maastrichtian–Danian boundary.

?Family Megathyrididae Dall 1870

Genus *Gwyniella* n.gen.

Pl. 11:1–5; Fig. 19A–D

Derivation of the name. – After the similarity to the genus *Gwynia* King 1859.
Type and only species. – *Gwyniella persica* n.sp.

Diagnosis. – Micromorphic megathyridid(?) with subcircular to subpentagonal outline, resupinate to biconvex shell; very low and recessed beak; foramen hypothyridid. Hinge very weak. Brachidium poorly developed, consisting of short, delicate crura and short, high and pointed median septum in dorsal valve. Lophophore probably schizolophous.

Remarks. – The genus *Gwyniella* exhibits very few characteristic features apart from the delicate median septum, and therefore it is uncertain whether it belongs to the family Megathyrididae. The similarity to the genus *Gwynia* King is that both genera are extremely neotenic; they are both minute and smooth-shelled. *Gwynia* is, however, almost linguloid in outline, trocholophos and only known from the Pleistocene to Recent (Fisher & Oehlert 1891; Hatai 1965).

The genera of the family Platidiidae Thomson 1927 (*Platidia*, *Amphithyris*, *Aemula* and *Scumulus*) differ from *Gwyniella* gen.n. in having amphithyridid foramen and a well-developed median septum, and in being spirolophous.

Gwyniella persica n.sp.
Pl. 11:1–5; Fig. 19A–D

Derivation of the name. – Latin *persica*, a peach; after the rounded, smooth shape.

Holotype. – MGUH No. 16953, Pl. 11:1; Lower Danian, Nye Kløv, Sample NK26. Type locality: Nye Kløv, north of Thisted, northwestern Denmark. Type horizon: Lower Danian.

Material. – Seventy-five complete shells, 19 dorsal and 38 ventral valves and a number of fragments. The material is

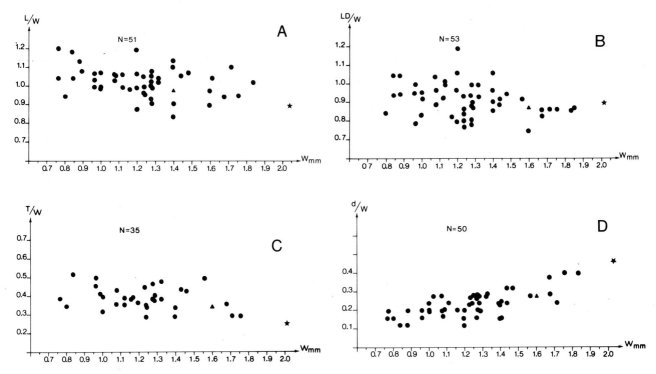

Fig. 19. Scatter diagrams of *Gwyniella persica* n. sp., Lower Danian, Nye Kløv. ▲: holotype, sample NK26, MGUH 16953; ★: largest specimen, sample NK22. □A. Ratio shell length L (mm) to width W (mm). □B. Ratio dorsal valve length LD (mm) to width W (mm). □C. Ratio thickness T (mm) to width W (mm). □D. Ratio foramen width d (mm) to width W (mm).

strongly recrystallized. The largest specimen is from the Lower Danian sample NK22 and has the following dimensions: length 2.12 mm; dorsal valve length 1.76 mm; width 2.32 mm; thickness 0.60 mm; width of foramen 0.48 mm. Measurements of the holotype: Length 1.56 mm; dorsal valve length 1.40 mm; width 1.60 mm; thickness 0.56 mm; width of foramen 0.28 mm.

Diagnosis. – Small, biconvex, smooth, thin shell with subcircular to semicircular outline. Beak very low, suberect; foramen hypothyridid, low and broad. Hinge and brachidium weakly developed; brachidium consists of short crura and high, short and pointed dorsal median septum.

Description. – The shell is minute and thin; the outline is subcircular to subpentagonal. In juveniles, the outline is oval to subcircular; in the later stages, the width exceeds the length, and an almost semicircular outline is developed. The maximum width is around the mid-length of the shell, and the auricles are indistinct. The shell is flat biconvex to resupinate, and the dorsal valve is of the highest convexity. The early growth stages are always biconvex. The front commissure is rectimarginate to incipient uniplicate. The shell is smooth, and the growth lines are seldom seen, and then only at the shell margin. The hinge line is broad. The beak is very low, recessed to suberect. The area is narrow and distinctly limited, and the foramen is hypothyridid, low, broad and triangular to trapezoidal. The foramen is acute in the early growth stages but often attrite in the later stages. The deltidial plates are narrow and most frequently disjunct posteriorly. The pedicle collar is well developed. The cardinal process is very small. The hinge is poorly developed. The hinge teeth are short and blunt, and dental plates are not

seen. The inner sockets are equally low and the inner socket ridges are low, very short and almost parallel. The brachidium consists of very short and slightly ventrally directed crura anterior to the inner socket ridges, and a short, high and pointed median septum placed anterior to the midlength of the dorsal valve. The material is heavily recrystallized, and most of the original shell material has been replaced by large calcite crystals. L/W, LD/W, T/W and d/W for the species are shown in Fig. 19A–D.

Remarks. – *Gwyniella persica* n.sp. is an example of a brachiopod with almost no characteristica, and is highly neotenic in its character.

Occurrence. – *Gwyniella persica* n.sp. occurs in the Danian samples NK17, NK18, NK19, NK21, NK22, NK24, NK25, NK26, NK27, and NK28, as well as in one sample (called L113) from 4.50 m above the boundary. It is typically represented by only a small number of individuals. In some of the most benthos-rich levels (samples NK22, NK25, NK26), however, the species is abundant.

Genus *Argyrotheca* Dall 1900

Type species. – *Terebratula cuneata* Risso 1826, by original designation.

Argyrotheca bronnii (Roemer 1841)

Synonymy. – □ 1841 *Terebratulina bronnii* v. Hag. – Roemer, p. 41, no. 31. □ 1842 *Orthis buchii* n. – Hagenow, p. 544, Pl. 9:8a–d. □ 1894 *Argiope bronnii* v. Hag. sp. – Ravn, p. 377, no. 37; p. 390, no. 37. □ 1909 *Argiope bronnii* v. Hagenow –

Nielsen, p. 171, no. 39. ☐ 1965 *Argyrotheca bronnii* (Roemer) – Steinich, pp. 124–134, Figs. 179–196, Pl. 17:1a–d, 2a–d. ☐ ?1965 *Argyrotheca lacunosa* sp.n. – Steinich, pp. 134–137, Figs. 197–199, Pl. 18:1a–d. ☐ 1972 *Argyrotheca bronnii* (Roemer) – Surlyk, p. 20. Figs. 5, 12, 14, 15, 16, 17, 18, Pl. 4g. ☐ 1979 *Argyrotheca bronnii* (Roemer) –Bitner & Pisera, pp. 78–79, Pl. 4:1–2. ☐ 1982 *Argyrotheca bronnii* (Roemer) – Surlyk, Fig. 1, Pl. 3i.j. ☐ 1984 *Argyrotheca bronnii* (Roemer) – Surlyk & Johansen, Fig. 1.

Material. – One complete shell, 20 dorsal and 18 ventral valves and a number of fragments. The largest specimen is from the Upper Maastrichtian sample NK1 and is represented by a broken dorsal valve with the following dimensions: length 3.7 mm; width 8.0 mm; width of foramen 1.5 mm; hinge line width 8.0 mm; and number of ribs 7.

Description. – The material from the Upper Maastrichtian of Nye Kløv is consistent with the descriptions of *Argyrotheca bronnii* given by Roemer (1841) and Steinich (1965). *A. bronnii* is characterized by its 6–10 low, radial ribs, which are only slightly extended beyond the shell margin, in its semicircular to subquadratic outline, in its low area, and in possessing disc-shaped, concave hinge plates and a strong median septum with a distinct cleavage.

Remarks. – In external features the Lower Danian *Argyrotheca* aff. *bronnii* is similar to *A. bronnii* from the Upper Maastrichtian, although the ribs of the Lower Danian specimens are more arched and protrude further beyond the shell margin. In internal features, the Lower Danian form differs in its hinge plates, and is thus referred to as *A.* aff. *bronnii*. Large specimens of *A. bronnii* continue morhologically into *A. coniuncta* Steinich, but the hinge plates of *A. coniuncta* differ in forming a coherent, distinct platform supported by the base of the septum, and in having a median septum which is not cleaved. *A. hirundo* (Hagenow) is distinguished from *A. bronnii* in possessing a smaller shell size, a higher area, fewer ribs (4–7), and in hinge plates fused into a narrow triangular platform.

Occurrence. – *A. bronnii* occurs in the Upper Maastrichtian samples NK1, NK4, NK5, NK6, NK7, NK8, and possibly NK9. The species is present in a small number of individuals and does most probably become extinct at the Maastrichtian/Danian boundary.

Argyrotheca aff. *bronnii* (Roemer 1841)
Pls. 11:6, 7; 12:1; Fig. 20A–E

Synonymy. – ☐ 1984 *Argyrotheca* n.sp.aff. *bronnii* – Surlyk & Johansen, Fig. 1.

Material. – Two complete shells, 18 dorsal and two ventral valves and a number of fragments. The largest specimen is represented by a dorsal valve from the Lower Danian sample NK23. The specimen has the following dimensions: length 3.2 mm; width 5.4 mm; width of foramen 0.9 mm; hinge line width 5.4 mm; and number of ribs 6.

Description. – This species is characterized by its often almost quadrangular dorsal valve. The shell surface contains 6–10 straight and rather prominent radial ribs which may protrude well beyond the shell margin. The foramen is high, pointed subtriangular and delimited by narrow ridge-like deltidial plates and a relatively wide area. The inner socket ridges are long and parallel to the posterior shell margin. In small specimens the hinge plates form two irregularly delimited plates, and in larger forms fuse into a deep and relatively wide concave disc.

Remarks. – Externally the species is similar to *Argyrotheca bronnii* (Roemer), although its ribs may appear more prominent and the shell more biconvex. Internally, however, *A. bronnii* differs in possessing hinge plates that are always formed as two separate concave discs. It is uncertain whether the Lower Danian specimens represents a new species or a variety of *A. bronnii*, but because of the morphological similarities the specimens are here referred to as *Argyrotheca* aff. *bronnii*. For comparison the Lower Danian *A.* aff. *bronnii* are shown in scatter diagrams for *A. bronnii* from the Lower Maastrichtian of Rügen (Fig. 20A–E).

Occurrence. – The species occurs in the Lower Danian samples NK17, NK19, NK21 and NK23.

Argyrotheca coniuncta Steinich 1965
Pl. 12:5A, B

Synonymy. – ☐ 1965 *Argyrotheca coniuncta* n.sp. – Steinich, pp. 138–144, Figs. 200–206, Pl. 18:2a–d. ☐ 1969 *Argyrotheca coniuncta* Steinich – Surlyk, pp. 194–199, Figs. 206–212, Pl. 20:1–3. ☐ 1972 *Argyrotheca coniuncta* Steinich – Surlyk, p. 20, Figs. 5–12. ☐ 1979 *Argyrotheca coniuncta* Steinich – Bitner & Pisera, p. 79, Pl. 5:5–7. ☐ 1984 *Argyrotheca coniuncta* Steinich – Surlyk & Johansen, Fig. 1.

Material. – Four dorsal valves and one ventral valve. The largest specimen is represented by a ventral valve from the Upper Maastrichtian Sample NK8 and has the following dimensions: length 2.0 mm; width = width of hinge line 4.0 mm; foramen width 1.4 mm; number of ribs 6.

Description. – The species conforms with the description of *Argyrotheca coniuncta* Steinich by Steinich (1965). *A. coniuncta* has a semicircular outline, very often much wider than long with the hinge line characteristically prolonged into winglike extensions. The anterior commissure is straight, and 6–11 low ribs are present on the shell surface. The beak is erect and the angle between the area and the commissure is characteristically large. The foramen is very large, hypothyridid and triangular, and the hinge is well developed. The anterior margins of the inner socket ridges are much longer than the posterior margins. The cardinal process is small. The hinge plates are fused mid-dorsally into a platform. The dorsal median septum is high and strong; both the anterior and the posterior edges of the septum curve in an arc towards the dorsal valve floor. The septum itself is not split, but the fusion of the septum and the descending branches of the brachidium give the median septum a characteristically split appearance. The lophophore was a schizolophe.

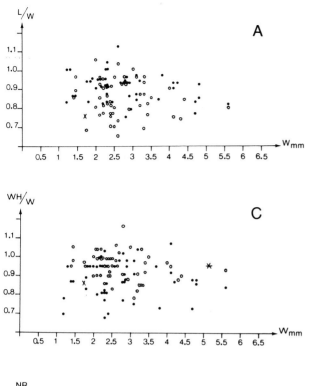

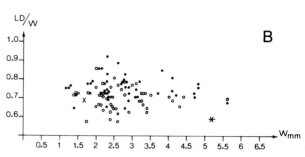

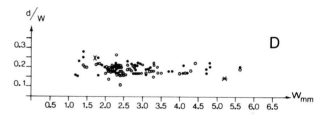

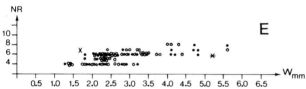

Fig. 20. Scatter diagrams of *Argyrotheca* aff. *bronnii*, Lower Danian, Nye Kløv (×; ★: largest specimen, sample NK23), and *A. bronnii* (Roemer), Lower Maastrichtian, Rügen (○, ●; data from Steinich 1965, Figs. 179, 180, 182, 185, 192. □A. Ratio shell length L (mm) to width W (mm). □B. Ratio dorsal valve length LD (mm) to width W (mm). □C. Ratio hinge line width WH (mm) to width (mm). □D. Ratio foramen width d (mm) to width W (mm). □E. Number of ribs NR.

Remarks. – Large forms of *Argyrotheca bronnii* (Roemer) differ from *A. coniuncta* in having a more rectangular outline and a cleaved median septum.

Occurrence. – *Argyrotheca coniuncta* occurs in the uppermost Maastrichtian samples NK7 and NK8 in a low number of individuals. The species becomes extinct at the Maastrichtian–Danian boundary.

Argyrotheca danica (de Morgan 1883)
Pl. 12:2–4

Synonymy. – □ 1833 *Cistella danica* – de Morgan, p. 394, Pl. 12:27–30. □ 1894 *Argiope danica* (de Morgan) – Posselt, p. 49. □ 1909 *Argiope danica* (de Morgan) – Nielsen, pp. 171–172, Pl. 1:43–45. □ cf. 1928 *Argiope buchii* v.Hag. – Nielsen, p. 218, Pl. 4:4–6. □ 1965 *Argyrotheca danica* (de Morgan) – Steinich, p. 145. □ 1984 *Argyrotheca danica* (de Morgan) – Surlyk & Johansen, Fig. 1.

Material. – One complete shell, one dorsal and two ventral valves. The largest specimen is represented by a ventral valve from the Upper Maastrichtian sample NK4 and has the following dimensions: length 1.6 mm; dorsal valve length 1.5 mm; width 2.4 mm; foramen width 0.5 mm; and number of ribs 4.

Description. – *Argyrotheca danica* has characteristically four extended ribs, a frontal incision of the shell margin, a flat triangular cardinal process and hinge plates formed as two concave, rounded discs anterior to the cardinal process. The shell is small, the outline is subpentangular, and the shell is clearly wider than long. The maximum width lies at the hinge line, which is markedly prolonged into wing-like extensions. The shell is flat biconvex, and the anterior commissure is straight. The four ribs are wide, flat, smooth and straight, and extend well beyond the shell margin. A deep, smooth median sinus is developed early in ontogeny and gives the shell a characteristic frontal incision. Three to four distinct growth lines are seen. The hinge line is very long and straight, the beak is acute and erect, and the area is relatively long and narrow. The foramen is large, hypothyridid and triangular, and the deltidial plates are narrow ridges, the height of which diminishes towards the apex. The pedicle collar is well developed. The inner socket ridges are low and diverge strongly anteriorly, and the anterior margins of the inner socket ridges are much longer than the posterior margins. The cardinal process is a flat, triangular body wedged in between the posterior margins of the inner socket ridges. The hinge plates are two separate, rounded, concave areas anterior to the cardinal process. The dorsal median septum reaches the ventral valve, and the septum in lateral view is a tapered triangular plate with a distinct groove in the posterior face. The posterior face slopes steeply towards the apex of the shell. The anterior face is almost perpendicular to the valve floor, and reaches almost to the incision in the shell margin. The inner surface of the shell is a negative relief of the outer surface.

Remarks. – Due to the very sparse material, it is difficult to establish clearly the criteria necessary to distinguish *Argyrotheca danica* from *A. bronnii* (Roemer). *A. bronnii* is distinguished from *A. danica* in possessing a larger number of ribs. At a shell width of 2.40 mm, *A. bronnii* possesses 6–7 ribs, whereas *A. danica* at the same size only has 4 ribs. But the shape of the hinge plates, the cardinal process and the dorsal septum are similarly developed. *A. obstinata* Steinich is distinguished from *A. danica* in having a higher area and being more biconvex. *A. hirundo* (Hagenow) differs from *A. danica* in its shorter hinge line, in the pivot-shaped cardinal process and the unsplit dorsal median septum.

Occurrence. – *Argyrotheca danica* is found in the Upper Maastrichtian samples NK4 and NK5 in a low number of individuals. It becomes extinct at the Maastrichtian–Danian boundary.

Argyrotheca stevensis (Nielsen 1928)
Pls. 12:6A, B; 13:1–4; Fig. 21A–F

Synonymy. – □ 1928 *Argiope stevensis* n.sp. – Nielsen, p. 219, Pl. 4:10–11. □ 1969 *Argyrotheca stevensis* (Nielsen) – Surlyk, pp. 206–210, Figs. 221–228, Pl. 13:5–7. □ 1972 *Argyrotheca stevensis* (Nielsen) – Surlyk, p. 20, Figs. 2, 5, 13. □ 1984 *Argyrotheca stevensis* (Nielsen) – Surlyk & Johansen, Fig. 1.

Material. – There are 762 complete shells, 135 dorsal and 164 ventral valves, and a number of fragments. The largest specimen is from the Lower Danian sample NK 24 and has following dimensions: length 2.7 mm; dorsal valve length 2.0 mm; width 2.1 mm; thickness 1.1 mm; width of foramen 0.8 mm; number of ribs 7; and width of hinge line 1.9 mm.

Description. – The shell is small, and the outline varies from subtriangular to subpentagonal. The lateral edges of the shell are often parallel. The maximum width varies from being placed at the hinge line to being placed anterior to the mid-length of the shell. The auricles are small. The shell is strongly biconvex with the ventral valve being most convex. The anterior commissure is straight. Four strong, straight and smooth radial ribs are formed at a shell width of about 0.80 mm. At a shell width of 1.20 mm, two new ribs are formed laterally to the rest, and the larger forms may contain two more ribs at each side. The ribs are straight and arranged in two almost parallel bundles separated by a median sinus. The median sinus is shallow and as wide as a rib, and a weak rib may be present in it. Four to five distinct growth lines are present. The hinge line is broad and straight and the beak is high, erect to slightly incurved and often attrite. The area is often concave, broad, and triangular due to the incurved nature of the beak. The foramen is high, large, hypothyridid and triangular. A pedicle collar is well developed. Laterally, the foramen is limited by two distinct deltidial plates, forming narrow ridges. The hinge is strong and the teeth are broad, strong and blunt. The inner socket ridges are short, relatively high, and diverge strongly anteriorly, and the anterior margins are somewhat prolonged laterally. The cardinal process is small and protruding, and the hinge plates are developed as two rounded, concave, plate-like bodies anterior to the cardinal process. The crura are either very small or missing; most frequently

it seems as if the descending branches unite with the inner socket ridges. The brachial bands are narrow and from a gentle curve approaching the dorsal valve floor, where they fuse with the floor around the midline of the shell. Further anteriorly, they are ventrally directed approaching the plate-like dorsal median septum with which the brachial bands fuse. The crural processes are short and grew in a medio-posterior direction. The dorsal median septum is strongly built and almost touching the floor of the ventral valve; the septum is a long, tapered triangular plate with the anterior face almost perpendicular to the shell floor. The posterior face slopes very sharply, and the top of the septum is anterior to the midline of the shell. The ventral median septum is low and long. The lophophore was probably a schizolope. The inner shell surface is a negative relief of the outer.

Ratios of morphological dimensions of *A. stevensis* from the Lower Danian of Nye Kløv are shown in scatter diagrams (Fig. 21A–F) together with those of Maastrichtian specimens measured by Surlyk (1969).

Remarks. – *A. stevensis* from the Danish Maastrichtian is found to have a generally higher WH/W ratio than the specimens from the Lower Danian (Fig. 21C), reflecting the more triangular outline of the Danian specimens. In other respects the morphology of the Maastrichtian form is similar to that of the Danian form. *Argyrotheca hirundo* (Hagenow) is distinguished from *A. stevensis* in being more flatly biconvex, in often having four shallow ribs which protrude distinctly beyond the anterior shell margin, and in having a very high area. *A. obstinata* Steinich is distinguished from *A. stevensis* in constantly having four ribs, in having a very high, almost plane area, and in being clearly elongated.

Occurrence. – *Argyrotheca stevensis* occurs in the Upper Maastrichtian samples NK3, NK5, NK8, and in the Lower Danian samples NK17, NK18, NK19, NK20, NK21, NK22, NK23, NK24, NK25, NK26, NK28.

A. stevensis is an index fossil for the uppermost Maastrichtian the *stevensis–chitoniformis* Zone of Surlyk (1982, 1984).

A. stevensis is found at Nye Kløv in the majority of the samples and in some samples from the upper part of the Lower Danian (samples NK24, NK25 and NK26) it dominates the brachiopod assemblage. There is a close relationship between abundance of bryozoans and number of individuals of *A. stevensis*. This is probably due to preference towards attachment to specific bryozoan species, a tendency which recent *Argyrotheca* spp. also possess (Ulla Asgaard, personal communication, 1983).

A. stevensis crosses the Maastrichtian–Danian boundary.

Argyrotheca aff. *stevensis* (Nielsen 1928)
Pl. 13:6, 7; Fig. 22A–F

Synonymy. – □ 1984 *Argyrotheca* n.sp.aff. *stevensis* (Nielsen) – Surlyk & Johansen, Fig. 1.

Material. – At least five complete shells. The largest specimen is from the Lower Danian sample NK23 and has the following dimensions: length 1.9 mm; dorsal valve length 1.3 mm; width 2.2 mm; thickness 0.8 mm; width of foramen 0.4 mm; width of hinge line 1.8 mm; and number of ribs 7.

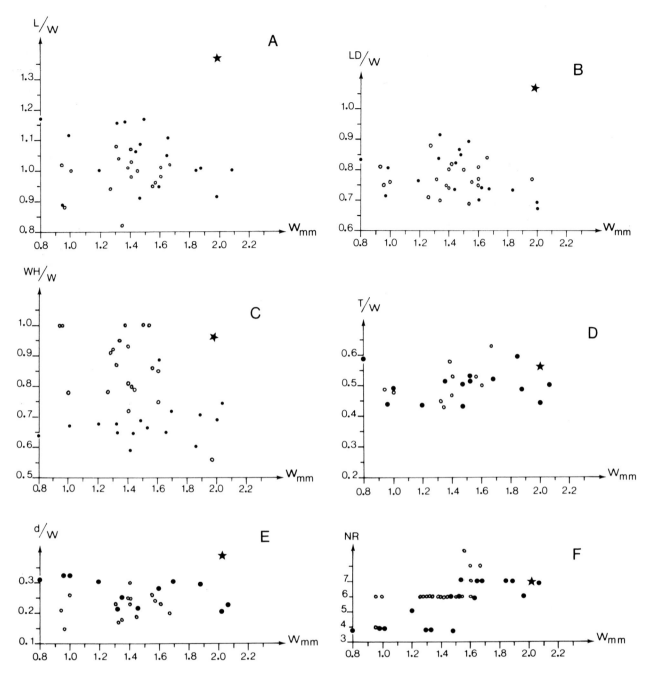

Fig. 21. Scatter diagrams of *Argyrotheca stevensis* (Nielsen), Lower Danian, Nye Kløv (●; ★: largest specimen, sample NK24) and *A. stevensis*, Maastrichtian, Denmark (○; data from Surlyk 1969, Figs. 221–224, 226–227). □A. Ratio shell length L (mm) to width W (mm). □B. Ratio dorsal valve length LD (mm) to width W (mm). □C. Ratio width of hinge line WH (mm) to width W (mm). □D. Ratio thickness T (mm) to width W (mm). □E. Ratio foramen width d (mm) to width W (mm). □F. Number of ribs NR.

Description. – The outline of the species is semicircular to subquadratic, and the maximum width lies around the mid-length of the shell. The auricles are indistinct, and the shell is biconvex with valves of equal convexity. The anterior commissure is straight, and the shell surface possesses 2–3 distinct growth lines. The shell surface shows 4–7 straight, smooth ribs that in the juvenile stages run along the length of the shell. Later in ontogeny the ribs are more radially directed. The ribs are in cross section high acute ridges, with narrow and deep interspaces. The ribs are extended markedly beyond the anterior shell margin, and are situated in two bundles on each side of a deep median sinus. A deep median incision is typically present in the anterior shell margin. The hinge line is extended into two wing-like processes, and the hinge line is straight and long, but does not exceed the width of the shell. In addition, there is a characteristic extension of the shell anterior to the wing-like processes. The beak is high, acute and erect, and the area is high, narrow and triangular. The foramen is large, hypothyridid and triangular and is restricted laterally by two narrow deltidial plates. The pedical collar is well developed. It has not been possible to investigate the interior morphology in detail, but the inner socket ridges are strong and markedly diverging anteriorly. The dorsal median septum is high and strong, and the anterior face of the septum reaches almost to the anterior shell margin. The posterior face reaches the mid-length of

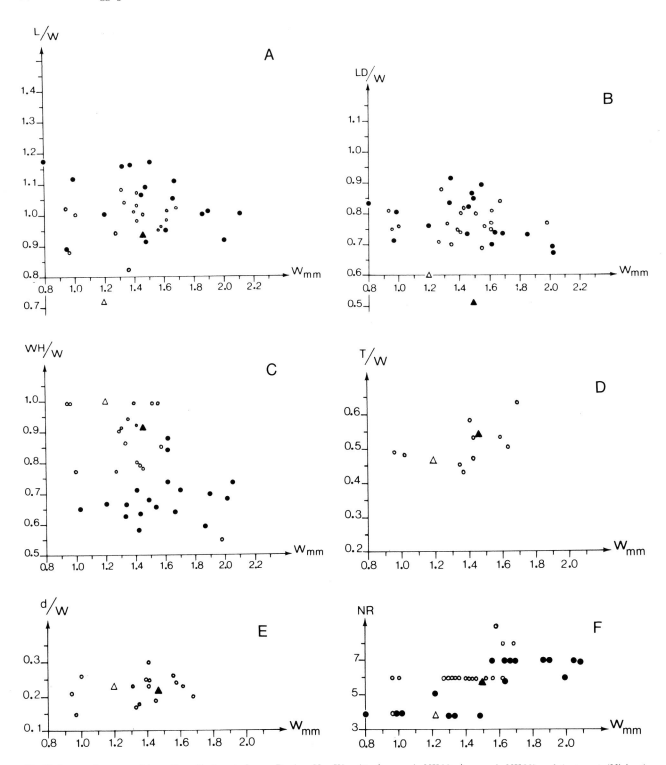

Fig. 22. Scatter diagrams of *Argyrotheca* aff. *stevensis*, Lower Danian, Nye Kløv (●; ▲: sample NK23; △: sample NK22) and *A. stevensis* (Nielsen), Maastrichtian, Denmark (○; data from Surlyk 1969, Figs. 221–224, 226–227). □A. Ratio shell length L (mm) to width W (mm). □B. Ratio dorsal valve length LD (mm) to width W (mm). □C. Ratio hinge line width WH (mm) to width W (mm). □D. Ratio thickness T (mm) to width W (mm). □E. Ratio foramen width d (mm) to width W (mm). □F. Number of ribs NR.

the shell. The lophophore was a schizolophe. The shell is relatively thin.

Remarks. – The form is consistent in many characters with the *Argyrotheca stevensis* (Nielsen) described herein. But it differs in the wing-like extensions of the hinge line and the characteristic wing-like extension anterior to the hinge line. *A. stevensis* is, however, rather variable and the specimens

might be varieties of *A. stevensis*. Further study of a larger collections may show the form to represent a new species or subspecies of *A. stevensis*. L/W, LD/W, T/W, NR/W, d/w and WH/W for the specimens are shown for comparison in the scatter diagrams (Fig. 22A–F) for *A. stevensis* from the Maastrichtian measured by Surlyk (1969).

Occurrence. – *A.* aff. *stevensis* occurs in the Lower Danian samples NK22, NK23, and possibly NK25.

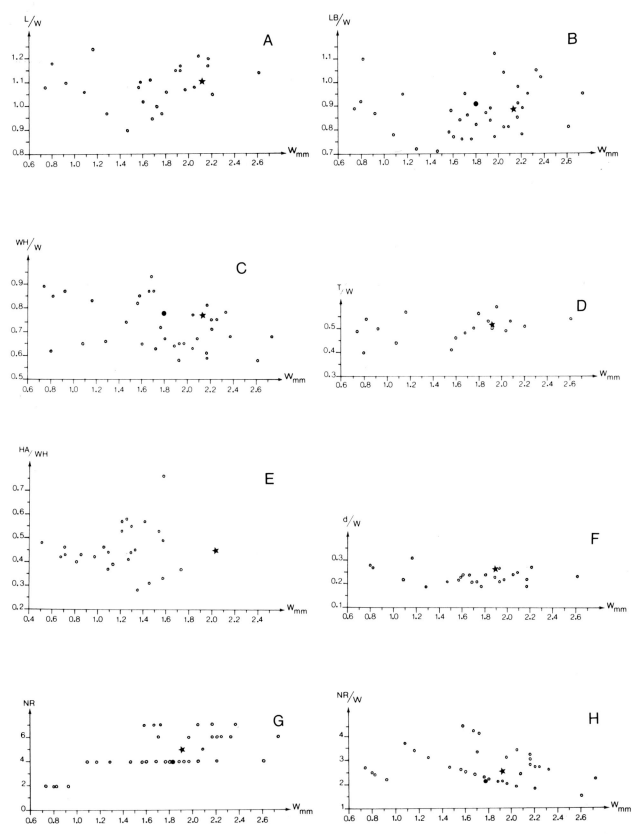

Fig. 23. Scatter diagrams of *Argyrotheca hirundo* (Hagenow) from the Upper Maastrichtian (●: largest specimen, sample NK4) and the Lower Danian, Nye Kløv (★: largest specimen, sample NK26), and from the Maastrichtian, Denmark (○; data from Surlyk 1969, Figs. 213–220). □ A. Ratio shell length L (mm) to width W (mm). □ B. Ratio length of dorsal valve LD (mm) to width W (mm). □ C. Ratio hinge line width WH (mm) to width W (mm). □ D. Ratio thickness T (mm) to width W (mm). □ E. Ratio area height HA (mm) to hinge line width WH. □ F. Ratio foramen width d (mm) to width W (mm). □ G. Number of ribs NR. □ H. Ratio number of ribs NR to width W (mm).

Argyrotheca hirundo (Hagenow 1842)
Pl. 14:1–8; Fig. 23A–H

Synonymy. – □ 1842 *Orthis hirundo* n. – Hagenow, p. 545, Pl. 9:9a–d. □ 1928 *Argiope hirundo* v. Hag. – Nielsen, pp. 218–219, Pl. 4:7–9. □ 1965 *Argyrotheca hirundo* (Hagenow) – Steinich, pp. 152–159, Figs. 232–249, Pl. 16:3, Pl. 17:3a–d. ?□ 1965 *Argyrotheca obstinata* sp.n. – Steinich, pp. 145–151, Figs. 217–231, Pl. 18:3a–d. □ 1969 *Argyrotheca hirundo* (Hagenow) – Surlyk, pp. 200–204, Figs. 213–220, Pl. 13:8–10. □ 1972 *Argyrotheca hirundo* (Hagenow) – Surlyk, p. 20, Figs. 5, 12, 15b, 17, 18. ?□ 1972 *Argyrotheca obstinata* (Hagenow) – Surlyk, p. 20, Figs. 5, 15b. □ 1979 *Argyrotheca hirundo* (Hagenow) – Bitner & Pisera, p. 79, Pl. 4:7. □ 1979 *Argyrotheca hirundo* (Hagenow) – Surlyk, Fig. 1, Pl. 3f, g, h. □ 1984 *Argyrotheca hirundo/Argyrotheca* aff. *hirundo* (Hagenow) – Surlyk & Johansen, Fig. 1.

Material. – Fifty-three complete shells, 13 dorsal and 11 ventral valves, and a number of fragments. The largest specimen is from the Lower Danian sample NK26 and has the following dimensions: length 2.1 mm; dorsal valve length 1.4 mm; width 1.9 mm; thickness 0.9 mm; width of foramen 0.5 mm; width of hinge line 1.5 mm; and number of ribs 5.

Description. – The shell is small, and the outline is longer than wide, subtriangular to subpentangular. The maximum width is at the anterior shell margin. The auricles are indistinct. The shell is biconvex with the ventral valve being much more convex than the dorsal, and the anterior commissure is straight to broadly sulcate. Four to seven strong, straight, smooth ribs are arranged in two bundles on each side of a wide and deep median groove on the shell surface. The ribs are in cross section rather acute ridges. Up to at least 14 distinct growth lines are seen. The hinge is relatively long and straight, and the beak is high, acute and erect to slightly incurved. The area is very high, planal and triangular, and the foramen is large, triangular and hypothyridid. The deltidial plates form narrow ridges, and a pedicle collar is well developed. The hinge and the inner socket ridges are strongly built. The cardinal process is a small and pivot-shaped body placed between the posterior margins of the inner socket ridges. The hinge plates are fused into a triangular, narrow, concave platform, or may form two closely spaced narrow triangular areas. The brachidium is well developed, and the crura are very short and emerge from the anterior margin of the inner socket ridges. The crural processes are short and converge mid-dorsally, and the descending branches are slender bands with small teeth on their anterior margin. The descending branches fuse with the valve floor approximately at the midline of the shell. The brachial bands rise from the valve floor and fuse with the median septum anterior to the mid-length of the shell. The dorsal median septum is uncleaved and forms a high, triangular plate. Its anterior edge reaches the ventral valve and is perpendicular to the valve floor. The posterior edge falls in a gentle concave arc towards the apex. The ventral median septum is very thin posteriorly and reaches the pedicle collar. The ventral septum is broader anteriorly and contains a shallow groove. The lophophore was a schizolophe. The shell is relatively thin.

Remarks. – *Argyrotheca hirundo* is a very variable species but is characterized by its pivot-shaped cardinal process and a rather constant, low rib number (4–7). The Lower Danian representations of this species differs from the Maastrichtian in possessing fewer ribs. For comparison the largest specimen of *A. hirundo* from the Upper Maastrichtian and the largest specimen from the Lower Danian are shown in the scatter diagrams for *A. hirundo* from the Danish Maastrichtian Chalk (Fig. 23A–H). The specimens from Nye Kløv fall within the area of variation of *A. hirundo*.

A. obstinata Steinich is distinguished from *A. hirundo* in having a higher area, in a more biconvex shell, in always possessing four ribs, and in lacking the pivot-shaped cardinal process. There is, however, a gradual transition between the two species, according to Surlyk (1969, p. 204).

A. stevensis (Nielsen) differs from *A. hirundo* in having 6–8 ribs arranged in bundles on each side of a median sinus. Within the bundles the ribs are parallel. Furthermore, the hinge plates in *A. stevensis* are developed as two rounded concave, disc-like bodies.

Occurrence. – *Argyrotheca hirundo* occurs in the Upper Maastrichtian samples NK4 and NK5 and in the Lower Danian samples NK21, NK22, NK23, NK25 and NK26. It is represented by a relatively high number of individuals.

Argyrotheca dorsata (Nielsen 1928)
Pl. 15:1, 2; Fig. 24A–G

Synonymy. – □ 1928 *Argiope dorsata* n.sp. – Nielsen, p. 216–217, 221–222, Pl. 5:25–27. □ 1968 *Argyrotheca dorsata* (Nielsen) – Asgaard, p. 115. □ 1984 *Argyrotheca* n.sp. 1 – Surlyk & Johansen, Fig. 1.

Material. – Nineteen complete shells, one dorsal and five ventral valves. The largest specimen is represented by a ventral valve from the Lower Danian sample NK26 and has the following dimensions: length 2.28 mm; width of foramen 0.56 mm; width of hinge line 1.36 mm; and number of ribs 11.

Description. – The shell is small, rather thick, subtriangular to subpentangular in outline. The auricles are indistinct and the shell is strongly biconvex. There are 7–11 smooth radial ribs, arranged on each side of a deep median sinus in which an intercalated rib is present. The ribs protrude well beyond the anterior shell margin. The hinge line is long and straight, and the beak is high, blunt and erect to slightly incurved. The area is long, concave, triangular, and the foramen is large and hypothyridid. The deltidial plates are narrow, well defined ridges. The pedicle collar and the hinge are both well developed. The inner socket ridges are relatively high, short, diverging anteriorly, and the cardinal process is small and bulbous. The hinge plates are fused into a concave, rectangular platform. The short and broad crura develop at the anterior margin of the inner socket ridges. The descending branches rise from the dorsal valve floor about two-thirds the length of the valve to fuse with the median septum. The dorsal median septum is a relatively low trapezoidal plate with a cleavage in the posterior ridge. The anterior edge of the septum plunges steeply towards the

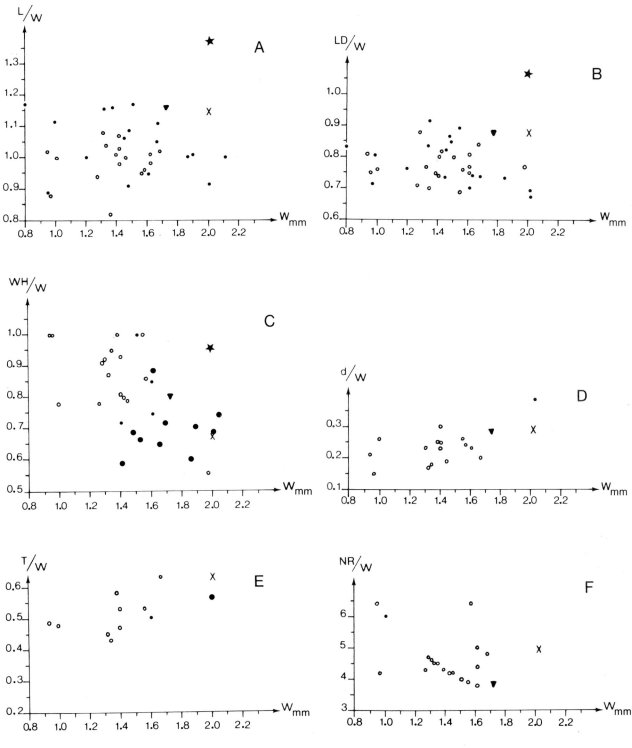

Fig. 24. Scatter diagrams of *Argyrotheca dorsata* (Nielsen), Lower Danian, Nye Kløv (×: largest specimen, sample NK26), and Upper Middle Danian, Rejstrup I, Nyborg, Fyn, Denmark (▼: holotype, Geological Museum Collection, Copenhagen, registration number MGUH 16873), and *A. stevensis* (Nielsen), Upper Maastrichtian, Denmark (○; data from Surlyk 1969, Figs. 221–224, 226–228) and Lower Danian, Nye Kløv (●; ★: largest specimen, sample NK24). □A. Ratio shell length L (mm) to width W (mm). □B. Ratio dorsal valve length LD to width W (mm). □C. Ratio hinge line width W (mm) to width W (mm). □D. Ratio foramen width d (mm) to width W (mm). □E. Ratio thickness T (mm) to width W (mm). □F. Ratio number of ribs NR to width W (mm). □G. Number of ribs NR.

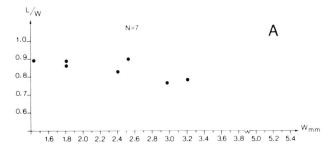

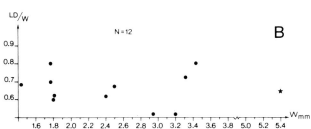

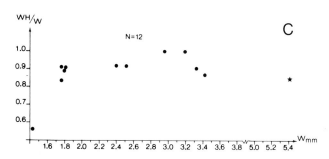

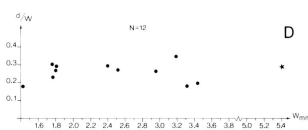

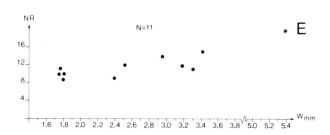

Fig. 25. Scatter diagrams of *Argyrotheca vonkoeneni* (Nielsen), Lower Danian, Nye Kløv. ★: largest specimen, sample NK29. ☐ A. Ratio shell length L (mm) to width W (mm). ☐ B. Ratio dorsal valve length LD (mm). ☐ C. Ratio hinge line width WH (mm). ☐ D. Ratio foramen width d (mm) to width W (mm). ☐ E. Number of ribs NR.

valve floor, and the posterior edge rather gently towards the apex. The ventral median septum is long, thin and relatively high. The lophophore was a schizolophe.

Remarks. – The holotype of *Argyrotheca dorsata* (Nielsen) is housed at the Geological Museum Collection of Copenhagen, registration number MGUH 16873. The type locality is the chalk quarry Rejstrup I, near Nyborg, Fyn, Denmark, and the holotype was found in Upper Danian bryozoan chalk (Ødum 1926; Nielsen 1928, Pl. 5:25–27). The holotype has the following dimensions: length 2.00 mm; length of dorsal valve 1.50 mm; width 1.72 mm; thickness 1.60 mm; width of foramen 0.48 mm; width of hinge line 1.40 mm; and number of ribs 7.

Although the internal morphology of the holotype is unknown, the specimens from Nye Kløv are similar to *A. dorsata* in their exterior features. The specimens are thus with slight reservation referred to this species.

A. dorsata is very difficult to separate from *A. stevensis* (Nielsen) but the latter differs in a lower number of ribs, an often subtriangular outline, a dorsal median septum developed as a long, high tapered triangular plate, and hinge plates developed as two concave discs.

For comparison, the holotype of *A. dorsata* and the largest specimen of *A. dorsata* from Nye Kløv are plotted into the scatter diagrams of *A. stevensis* (Fig. 24A–G).

A. hirundo (Hagenow) is distinguished from *A. dorsata* in

having a rather constant rib number (4), in possessing a much more flattened shell and in possessing a very high area. *A.* cf. *faxensis* differs from *A. dorsata* in possessing 5–9 well defined ribs arranged in a fan-like fashion and a dorsal median septum formed as a very high trapezoidal plate.

Occurrence. – *Argyrotheca dorsata* occurs in the Lower Danian samples NK21, NK22, NK24, NK26, and NK28. It is always present in a low number of individuals. This is the first recorded occurrence of the species in the Lower Danian.

Argyrotheca vonkoeneni (Nielsen 1928)
Pls. 16:1–5; 17:5; Fig. 25A–E

Synonymy. – ☐ 1928 *Argiope v.Koeneni* n.sp. – Nielsen, pp. 220–221, Pl. 4:22–24. ☐ 1984 *Argyrotheca* n.sp. 2 – Surlyk & Johansen, Fig. 1.

Material. – Five complete shells, 25 dorsal and 16 ventral valves, and a number of fragments. The largest specimen is a fragmentary shell from the Lower Danian sample NK29. It has the following dimensions: dorsal valve length 5.4 mm; width 4.6 mm; thickness 1.2 mm; width of foramen 1.6 mm; and number of ribs 20.

Description. – The species is relatively large for the genus. The outline is subpentagonal to pointed semicircular. The

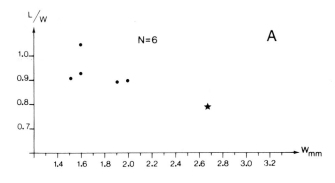

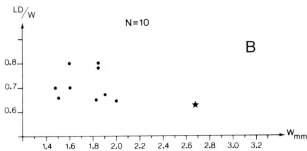

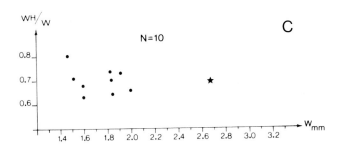

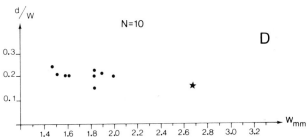

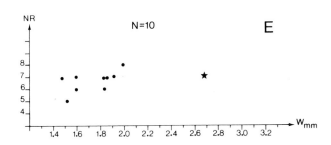

Fig. 26. Scatter diagrams of *Argyrotheca* cf. *faxensis* (Posselt), Lower Danian, Nye Kløv. ★: largest specimen, sample NK26. □A. Ratio length of shell L (mm) to width W (mm). □B. Ratio dorsal valve length LD (mm) to width W (mm). □C. Ratio hinge line width WH (mm) to width W (mm). □D. Ratio foramen width d (mm) to width W (mm). □E. Number of ribs NR.

maximum width of the shell occurs at its mid-length or at the hinge line, which is not alate as in many other species of *Argyrotheca*. The L/W, LD/W, WH/W, d/W and NR ratios are shown in Fig. 25A–E. The width of the shell is always somewhat greater than its length. The shell is flatly biconvex with an almost plane dorsal valve, and the anterior commissure is sulcate. New ribs are characteristically formed by intercalation, and the ribs protrude only slightly beyond the shell margin. Six smooth, straight narrow radial ribs are present at a shell width of 1.50 mm, at 3.20 mm there are 15 ribs, and larger specimens have as many as 20 ribs. The rib width is almost constant throughout ontogeny. The hinge line is long and straight. The beak is suberect, long, acute but often attrite, and the area is broad, triangular, and clearly delimited. The foramen is large, hypothyridid, subtriangular and confined laterally by narrow, plate-like deltidial plates. A pedicle collar with an irregular anterior margin is well developed. The hinge and the brachidium are well developed. The inner socket ridges are relatively low and short, and meet posteriorly at an acute angle. The anterior margins of the inner socket ridges are strongly extended laterally, giving the shell its alate appearance. The cardinal process is a small conical knob that extends slightly beyond the posterior shell margin. The hinge teeth are long, blunt ridges. The hinge plates are developed as a deep,

rather narrow, concave platform with a median ridge representing the fusion of the two plates. The brachial bands develop from the anterior of the inner socket ridges, and crura are either lacking or very small. The brachial bands are broad and ring-shaped, and rise from the midline of the shell. The high, plate-like, dorsal median septum is as long as the length of the dorsal valve and has a very short ventral edge. The posterior edge of the septum slopes steeply in a convex arc towards the apex and forms the highest point of the median septum. The anterior edge slopes more gently in a convex arc towards the anterior shell margin, and the septum is cleaved. The high, thin, ventral median septum reaches the midline of the shell. The lophophore was probably a schizolophe. The inner surface of the shell is smooth.

Remarks. – *Argyrotheca vonkoeneni* is distinguished from other species of *Argyrotheca* by its large number of intercalated ribs. In material from the Guelhem Chalk, Limburg, which is of Middle Danian age (Rasmussen 1965; Deroo 1966; Felder 1975), specimens with close affinities to *Argyrotheca vonkoeneni* are present. The specimens from Limburg contain 15–20 somewhat irregular ribs formed by intercalation; the outline, hinge and cardinalia are vary similar to that of *A. vonkoeneni*. The specimens from Limburg are, however, generally larger than the Danish specimens.

Occurrence. – *Argyrotheca vonkoeneni* occurs in the Lower Danian samples NK21, NK22, NK24?, NK29 and NK30 in a low number of individuals. In the lower Danian samples NK17 and NK19 there are fragments of a large *Argyrotheca* with many narrow, straight ribs, only slightly protruding from the shell margin. These may also belong to *A. vonkoeneni*, but the material does not allow for any confident assignment. This is the first recorded occurrence of the species in the Lower Danian of Denmark.

Argyrotheca cf. *faxensis* (Posselt 1894)
Pls. 15:3–6; 16:1; Fig. 26A–E

Synonymy. – □ 1894 *Argiope faxensis* n.sp. – Posselt, p. 52, Pl. 1:16, 17. □ 1928 *Argiope faxensis* Posselt – Nielsen, pp. 216–219, 225, Pl. 5:43–45. □ 1984 *Argyrotheca* n.sp. 3 – Surlyk & Johansen, Fig. 1.

Material. – Seven complete shells, nine dorsal and four ventral valves. The largest specimen is from the Lower Danian sample NK26 and has the following dimensions: length 2.12 mm; length of dorsal valve 1.68 mm; width 2.68 mm; thickness 0.88 mm; width of foramen 0.40 mm; width of hinge line 1.84 mm; and number of ribs 9.

Description. – The shell is small, with its maximum width at mid-length. The auricles are indistinct. The shell is flat, biconvex, with a flat dorsal valve. The anterior commissure is broadly sulcate, and a wide, flat median sinus is present on the valves. The ratios L/W, LD/W, WH/W, d/W and NR are shown in Fig. 26A–E. Five to nine straight, smooth and low radial ribs which extend beyond the shell margin are present. The ribs are in cross section low, acute ridges and are characteristically spread fan-like from the umbo. The interspaces between the ribs are wide. A small single rib may be developed in the median sinus. The hinge line is relatively wide and straight. There are 4–5 distinct growth lines. The beak is long, acute, suberect to erect, and the area is long, triangular and slightly concave. The foramen is large, triangular, and hypothyridid, and laterally limited by two distinct narrow deltidial plates that may meet posteriorly. A pedicle collar is well developed and the teeth are short and blunt. The inner socket ridges are low and short and meet posteriorly at an acute angle. The anterior margins of the inner socket ridges are only slightly prolonged laterally. The cardinal process is a low triangular plate, and the hinge plates are fused into a narrow ridge anterior to the cardinal process. The crura are short and rise close to each other on the inner socket ridges. The brachial bands are broad with a serrate outer margin and rise from the valve floor anterior to the midline of the shell. The dorsal median septum is a long, high, trapezoidal plate; the anterior edge of the median septum slopes steeply towards the shell margin, and the posterior one in a gentle concave arc towards the apex. The ventral edge is relatively long. The dorsal median septum is cleaved. The ventral median septum is long and low. The lophophore was probably a schizolophe. Inner surface of the shell is a negative relief of the outer.

Remarks. – The holotype of *Argyrotheca faxensis* (Posselt) has unfortunately not been available for study, and comparison

of the specimens from Nye Kløv and the holotype is thus based only on Posselt's very short description and blurred illustrations (Posselt 1894, p. 52, Pl. 1:16, 17). The specimens from Nye Kløv are characterized by a broad subtriangular to broad pointed oval outline, 5–9 radial ribs arranged in a fan-like fashion, hinge plate formed as a fused, narrow ridge, and a dorsal median septum formed as a long, high, trapezoidal plate.

In their external features the specimens from Nye Kløv appear similar to the holotype, except for a slightly lower number of ribs. But as the internal morphology of the holotype is not known, the specimens from Nye Kløv are referred to as *Argyrotheca* cf. *faxensis*. Nielsen (1928) redescribed *A. faxensis*. His specimens are very similar to the ones from Nye Kløv (Nielsen 1928, pp. 216–219, 225, Pl. 5:43–45).

A. stevensis (Nielsen) differs from *A.* cf. *faxensis* in having a subtriangular outline, often wing-like extensions of the ribs, ribs arranged in two bundles on each side of a median sinus with the ribs parallel within the bundles, and hinge plates developed as two rounded concave discs.

A. dorsata (Nielsen) is distinguished from *A.* cf. *faxensis* by its subpentangular outline, strongly biconvex shell and dorsal median septum formed as a low trapezoidal plate.

In a coarse calcarenite from the Middle Danian Guelhem Chalk, Limburg (Rasmussen 1965; Deroo 1966; Felder 1975), an *Argyrotheca* species is present which shows strong affinities to *A. faxensis*. The specimens from Limburg are generally of larger size, but possess the same number of ribs and the same shape of the dorsal median septum.

Occurrence. – *Argyrotheca* cf. *faxensis* occurs in the Lower Danian samples NK26 and NK28. The species is present in only a small number of individuals. This is the first recorded occurrence of the species in the Lower Danian.

Argyrotheca armbrusti (Schloenbach 1866)
Pl. 17:2–4

Synonymy. – □ 1866 *Argiope armbrusti* sp. nov. – Schloenbach, pp. 48–49, Pl. 3:4–8. □ 1984 *Argyrotheca* sp. 4 – Surlyk & Johansen, Fig. 1.

Material. – Three complete shells, four dorsal and one ventral valve. The largest specimen is from the Lower Danian sample NK and has the following dimensions: length 2.96 mm; dorsal valve length 2.40 mm; width 4.56 mm; thickness 1.48 mm; width of foramen 0.96 mm; width of hinge line 4.56 mm; and number of ribs 15.

Description. – The shell is relatively large, with a broad, semicircular outline always wider than long. The hinge line is prolonged into prominant wing-like extensions, and the maximum shell width lies at the hinge line. The shell is biconvex with the ventral valve most convex, and the auricles are indistinct. The anterior commissure is broad and sulcate. There are 6–8 ribs at a shell width of 1.40 mm; 12 ribs at 3.20 mm and 15 ribs at a shell width of 4.5 mm. New ribs are formed by intercalation. At a shell width of 1.75 mm, an incipient median sinus is present. The medium sinus is wide and posseses a low, single rib. The ribs are relatively narrow, smooth, acute ridges, and the interspace between them is often as wide as the ribs. Only a few distinct growth lines are present. The hinge line is very long and straight, and the area is large, triangular and somewhat receded. In the juveniles the area is very narrow compared to the relative size of the adult area. The deltidial plates are narrow, disjunct ridges. The beak is very high and erect, the foramen is very large, hypothyridid and triangular, and the pedicle collar is broad and flat with a convex anterior margin. The hinge is weakly developed. The dorsal median septum is high and plate-like with the distal end reaching the ventral valve. The ventral edge of the septum is rather long, and the anterior face possesses a shallow groove and is almost perpendicular to the shell floor. The posterior edge slopes steeply towards the posterior margin of the shell. The ventral median septum is low and reaches the pedicle collar. The lophophore was probably a schizolophe, the inner surface of the shell is a weak negative relief of the outer surface. The shell is rather thick.

Remarks. – In their external features the specimens from Nye Kløv closely resemble *Argyrotheca armbrusti* (Schloenbach), especially in their strongly alate shells, intercalated ribs and very large foramen. Although only little is known about the internal features of the holotype, their external features are very characteristic and similar. The specimens described here are therefore referred to as *A. armbrusti*, although with slight reservation. *A. coniuncta* Steinich is distinguished from *A. armbrusti* in having fewer ribs, well developed hinge plates, and a dorsal median septum which does not posses a cleavage. *A. vonkoeneni* (Nielsen) resembles the specimens described here, especially in its large shell size and in possessing ribs formed by intercalation. However, *A. vonkoeneni* differs from *A. armbrusti* in lacking wing-like extensions of the hinge line and in having a larger number of ribs.

Occurrence. – The species is found in the Lower Danian samples NK17 and NK19. It occurs in a small number of individuals. This is the first recorded occurrence of the species in the Lower Danian.

Family Platidiidae Thomsen 1927

Genus *Platidia* Costa 1852

Type species. – *Orthis anomioides* Scacchi & Philippi 1844, by original designation.

Platidia sp.
Pl. 19:1–3

Synonymy. – ☐ 1984 *Platidia* sp. – Surlyk & Johansen, Fig. 1.

Material. – One complete shell, three dorsal and two ventral valves. The largest specimen is represented by a dorsal valve from the Lower Danian sample NK25 and has the following dimensions: length 2.20 mm; width 2.40 mm; width of foramen 0.44 mm.

Description. – The shell is small, and the outline is irregularly oval and pointed. The maximum width is anterior to the

mid-length. The shell is flatly biconvex with the ventral valve slightly more convex than the dorsal. The shell surface is smooth, and no growth lines can be seen. The hinge line is short and straight, and the beak is low, acute and suberect. The area is narrow and well defined, and the deltidial plates are two distinct, narrow short ridges that meet posteriorly. The foramen is large, irregular, amphithyridid, rounded, irregular or asymmetric, and reaches far into the dorsal valve. A short pedicle collar is developed. The hinge is relatively strongly built, and the hinge teeth are feeble. The inner socket ridges are short, blunt parallel ridges, and the inner sockets are deep. A cardinal process is present. Hinge plates are not developed. The material does not allow for a complete investigation of the brachidium. However, the inner socket ridges follow the arc of the foramen, and at the anterior end of the arc, the inner socket ridges rise from the valve floor and continue via the crura into two cylindrical, slender, long and ventrally converging descending branches. The median septum is a low and very thin plate, with the anterior edge as the highest point. The anterior edge slopes very steeply towards the valve floor, and the posterior edge slopes gently towards the foramen arc. The shell is thin. In the present material, the shells are strongly recrystallized into coarse, long and radially arranged calcite crystals.

Remarks. – The strongly amphithyridid foramen is very characteristic for the genera *Aemula* Steinich and *Platidia* Costa. Externally, the two genera are impossible to distinguish. But *Aemula* differs from *Platidia* in lacking crura and the descending branches, and in possessing hinge plates. The genus *Amphithyris* Thomson differs from *Platidia* in the form of the median septum and in having hinge plates. Bosquet (1859) and Schloenbach (1866) classified their *Morrissia* Davidson species as belonging to the genus *Platidia* Costa, but the species illustrated by Bosquet and Schloenbach have hypothyridid foramen, a larger beak and a differently built brachidium. Steinich (1968a) mentions that the majority of the *Morrissia* species described presumably belong to the genus *Scumulus* but, as will be discussed below, the morphological differences between *Scumulus* and *Morrissia* are so great that they most likely also belong to different genera.

Occurrence. – *Platidia* sp. occurs in the Lower Danian samples NK21, NK22 and NK25 in a low number of individuals.

Genus *Aemula* Steinich 1968

Type species. – *Aemula inusitata* Steinich 1968, by original designation.

Aemula inusitata Steinich 1968
Pl. 18:1–5

Synonymy. – ☐ 1968 *Aemula inusitata* sp.n. – Steinich, pp. 193–199, Figs. 1–5, Pl. 1:1. ☐ 1972 *Aemula inusitata* Steinich – Surlyk, pp. 20, 35, 40, Figs. 5, 12, 13, 15–18, Pl. 3e, f, h. ☐ 1974 *Aemula inusitata* Steinich – Surlyk, pp. 185–203, Figs. 1, 3, 6, 7, Pl. 2A, C–F. ☐ 1979 *Aemula inusitata* Steinich – Bitner & Pisera, pp. 79–80, Pl. 5:4. ☐ 1984 *Aemula inusitata* Steinich – Surlyk & Johansen, Fig. 1.

Material. – Five complete shells, eight dorsal and 34 ventral valves. The largest specimen is represented by a dorsal valve from the Upper Maastrichtian sample NK1 and has the following dimensions: length 2.88 mm; width 3.20 mm; width of foramen 0.64 mm.

Description. – The shell is small and thin, and the outline is oval to irregular subcircular; in larger specimens the shell is often wider than long. The shell is biconvex, plano-convex or concavo-convex. The dorsal valve is smooth with distinct growth lines, and the ventral valve is more or less covered with small tubercles. The beak is short and suberect, and the area is very small and distinctly delimited. The foramen is large, amphythyridid, irregular in outline, and limited by two narrow deltidial plates. The pedicle collar is short. The teeth are small and blunt. The hinge plates are low, distinct and ridge-formed, and crura are not present. The median septum is low and triangular with the ascending branches represented by two small lateral wings.

Remarks. – Although the Lower Danian material is sparse, the morphological similarities between this and the Maastrichtian specimens are so great that the Lower Danian specimens are referred to *Aemula inusitata* without reservation.

The genus *Platidia* Costa is reminiscent of *Aemula* in its external features, as both genera possess large, amphithyrid foramina and very low acute beaks. But in internal features *Platidia* differs from *Aemula* in possessing crura. *Amphithyris* Thomson differs from *Aemula* in possessing crura and descending branches.

Occurrence. – This species is found in the Upper Maastrichtian samples NK1, NK2, NK3, NK4, NK5, NK6, NK7, NK8 and NK10, and in the Lower Danian sample NK25.

Aemula inusitata is represented by a small number of individuals.

Genus *Scumulus* Steinich 1968

Type species. – *Scumulus inopinatus* Steinich, by original designation.

Scumulus inopinatus Steinich 1968
Pl. 18:8, 9; Fig. 27A–E

Synonymy. – ☐ 1968a *Scumulus inopinatus* n.sp. – Steinich, pp. 200–206, Figs. 6–9, Pl. 1:2. ☐ 1969 *Scumulus inopinatus* Steinich – Surlyk, pp. 215–218, Figs. 232–235, Pl. 19:1–4. ☐ 1972 *Scumulus inopinatus* Steinich – Surlyk, p. 20, Figs. 5, 12–16, 18. ☐ 1984 *Scumulus inopinatus* Steinich – Surlyk & Johansen, Fig. 1.

Material. – Thirty-five complete shells, four dorsal and six ventral valves. The largest specimen is from the Upper Maastrichtian sample NK1 and has the following dimensions: length 2.72 mm; dorsal valve length 2.32 mm; width 2.40 mm; thickness 0.72 mm; width of foramen 0.48 mm; angle of area-edge β 120°.

Diagnosis. – Small, smooth, thin shell, outline subtriangular to elongated oval. Shell is flattened biconvex, with short suberect to erect beak. Distinct pedicle collar often formed as a double arc anteriorly. Large, triangular amphithyridid to hypothyridid foramen. Brachidium represented by descending branches and narrow, short median septum. (Emended from Steinich 1968.)

Description. – The outline is subtriangular to elongated pointed oval, and the maximum width is situated anterior to the mid-length of the shell or, in juveniles, close to the anterior shell margin. The outline in larger forms is broad subtriangular to elongated oval. Surlyk (1969) mentions that outline of *Scumulus inopinatus* from Møns Klint often is irregular, but this is not the case with specimens from Nye Kløv. The auricles are indistinct. The shell is flattened biconvex to resupinate with the dorsal valve of highest convexity. The anterior commissure is rectimarginate to incipiently sulcate. The shell surface is smooth, and only a few distinct growth lines are observed.

The hinge line is short and weakly oblique. The beak is suberect to erect, and short and blunt. The area is broad and triangular. The foramen is large and in juveniles most frequently amphithyridid; in adults it is hypothyridid to weakly amphithyridid. The shape of the foramen is elongated triangular to subtriangular, and the deltidial plates are narrow but distinct. A pedicle collar is well developed, and its anterior margin is often formed as a double arc (Steinich 1968b, Figs. 7–3). The hinge is well developed, and the cardinalia reach far beyond the dorsal posterior shell margin. The hinge teeth are small, pointed, and posteriorly directed, and the inner sockets are shallow. Cardinal process, hinge plates and dental plates are not developed. The brachidium consists of short, ventrally directed crura, short, blunt, median and ventrally directed crural processes, and heavy descending branches. The latter are fused with a high, narrow median septum situated around the mid-length of the shell. The spicular skeleton is strongly developed, and the lophophore was probably a zygolophe.

Ratios of L/W, LD/W, T/W, and d/W are for comparison plotted in scatter diagrams for *S. inopinatus* constructed by Surlyk (1969; Fig. 27A–D herein). Area-edge angle is shown in Fig. 27E.

Remarks. – In general, the material from Nye Kløv has a larger variation in the area-edge angle than material from the Lower Maastrichtian of Rügen described by Steinich (1965). The largest specimen from Nye Kløv is markedly larger than the largest specimen in Surlyk's (1969) material from the Upper Maastrichtian of Møn.

The genus *Scumulus* shows some affinities to the *Morrissia* species described by Bosquet (1859) and Schloenbach (1866; see also Steinich 1968a, pp. 192, 193, 200). It is, however, questionable whether the species described by Schloenbach are congeneric or conspecific with specimens described by Bosquet, as Schloenbach has only given exterior views of the specimens.

Of the *Morrissia* species described by Schloenbach, the *Scumulus* species from Nye Kløv resembles most *Morrissia suessi* Bosquet and *Morrissia antiqua* Schloenbach (Schloenbach 1866, Pl. 39:16–17), but the Danish specimens are of

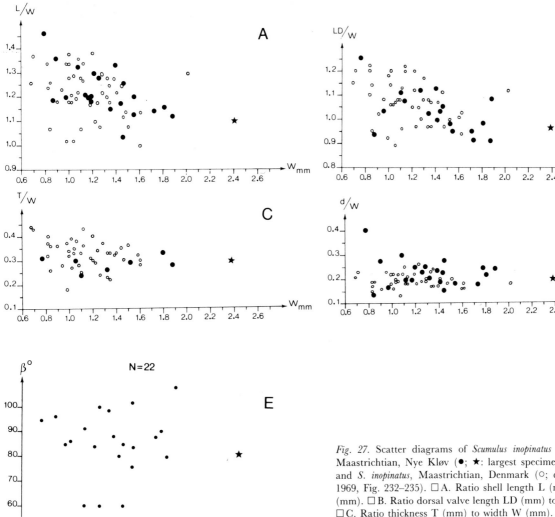

Fig. 27. Scatter diagrams of *Scumulus inopinatus* Steinich, Upper Maastrichtian, Nye Kløv (●; ★: largest specimen, sample NK1) and *S. inopinatus*, Maastrichtian, Denmark (○; data rom Surlyk 1969, Fig. 232–235). □A. Ratio shell length L (mm) to width W (mm). □B. Ratio dorsal valve length LD (mm) to width W (mm). □C. Ratio thickness T (mm) to width W (mm). □D. Ratio foramen width d (mm) to width W (mm). □E. Ratio area edge angle a (x) to width W (mm).

smaller shell size and do not show the fine radial ribs illustrated by Schloenbach. Of the *Morrissia* species described by Bosquet (1859), the material from Nye Kløv most resembles *Morrissia? suessi* Bosquet (Bosquet 1859, Pl. 5:15–18), but the morphological differences are here so prominent that the two forms most likely belong to different genera.

When analysing specimens of *Morrissia suessi* (Bosquet) from the Upper Maastrichtian and Danian of the Dutch Limburg area (St. Pietersburg), the distinction between *Scumulus* and *Morrissia* becomes clear. The specimens from Limburg are much larger than the specimens from Nye Kløv, and they possess fine but distinct radial ribs, a long, high median septum, dental plates, distinct cardinal process and hypothyridid foramen. Opposed to this, *Scumulus inopinatus* is smooth-shelled, small, has a short median septum, no cardinal process and no dental plates, and its foramen is in addition often amphithyridid.

A species that is referred to as *Scumulus?* sp. from the uppermost Maastrichtian of Nye Kløv is described below.

Occurrence. – The species is found in the Upper Maastrichtian samples NK1, NK3, NK4, NK5, NK6, NK7, and NK8. The species has not been found in the Danian. *S. inopinatus* is present as a low number of individuals.

Scumulus? sp.
Pl. 19:6–7

Synonymy. – □ 1984 *Scumulus(?)* sp. – Surlyk & Johansen, Fig. 1.

Material. – Ten complete shells and one ventral valve. The largest specimen is from the Upper Maastrichtian sample NK6 and has the following dimensions: length 1.84 mm; width 2.08 mm; dorsal valve length 1.44 mm; thickness 0.44 mm; and foramen length 0.40 mm.

Description. – The shell is small and thin, and the outline changes from pointed wide oval in juveniles to subpentangular later in the ontogeny. The maximum width changes during growth from around mid-length of the shell to the anterior of the shell. The auricles are indistinct. The shell is flat biconvex and in juveniles almost plano-convex. The dorsal valve stays rather flat throughout the ontogeny. During growth a shallow median depression is formed, and the anterior commissure thus changes from straight to sulcate. The shell surface is smooth, and a few growth lines can be traced. The hinge line is long and straight. The beak is low and suberect to erect, acute in juveniles, but very truncated later in ontogeny. The area is narrow and triangular, and the deltidial plates are two narrow low ridges that limit the foramen laterally. The foramen is relatively large, triangular in the juveniles and trapezoidal in the later stages. The foramen is hypothyridid to slightly amphithyridid, and a pedicle collar is well developed and reaches around one-third the length of the foramen.

Although opened, the shells do not allow for any investigation of their interior, and the morphology of the brachidium and hinge of this species is therefore unknown. If a median septum is present, it is but small. A strong spicular skeleton with long filaments is present in a recrystallized form.

Remarks. – The generic assignment of these specimens is uncertain as its interior is unknown. The species most closely-resembles the genus *Scumulus* Steinich, as both forms have slightly amphithyridid foramens and flat, smooth, biconvex shells. The outline of the species is, however, more quadratic than in *Scumulus inopinatus*, and the anterior margin of its pedicle collar does not exhibit a double anterior margin, nor is it fused with the inner surface of the shell as is characteristic for *S. inopinatus*.

Of the *Morrissia* species described by Schloenbach (1866, Pl. 39:14–17) *Scumulus?* sp. is most reminiscent of his Fig. 16 a–c of *Morrissia suessi* Bosquet, but Schloenbach's species differs in having fine radial ribs on the ventral valve.

Occurrence. – *Scumulus?* sp. is found in the Upper Maastrichtian samples NK4, NK5, and NK6, and is not found in the Lower Danian. The species occurs in low numbers of individuals.

Family Dallinidae Beecher 1893

Subfamily Kingeninae Elliott 1948

Genus *Kingena* Davidson 1852

Type species. – *Terebratula lima* Defrance 1828, by original designation.

Kingena pentangulata (Woodward 1833)
Pl. 20:5, 6

Synonymy. – □ 1833 *Terebratula pentangulata* – Woodward, p. 49. Pl. 6, Fig. 10. □ 1847 *Terebratula hebertiana*, d'Orbigny, p. 108, Pl. 514:5–10. □ 1852 *Kingena lima* Defrance – Davidson, p. 42, Pl. 4:16, 18, 20. □ 1961 *Kingena lima* (Defrance) – Peake & Hancock, p. 320. □ 1968b *Kingena* sp. – Steinich, pp. 342–345, Fig. 4–5. □ 1970 *Kingena pentangulata* (Woodward) – Owen, pp. 64–67, Pl. 7:6, 7, Pl. 8:2–6. □ 1972 *Kingena pentangulata* (Woodward) – Surlyk, p. 21, Figs. 5, 17. ?□ 1979 *Kingena* sp. – Bitner & Pisera, p. 81, Pl. 5, Figs. 1–2. □ 1984 *Kingena pentangulata* (Woodward) – Surlyk & Johansen, Fig. 1.

Material. – Two complete shells, 13 dorsal and five ventral valves, and 40 fragments. The complete shells are all juveniles; the dorsal and ventral valves are recognisable by their

internal morphology, but the material is too fragmented to allow for any measurements of the largest specimen. Only a few measurements have been obtained. A juvenile dorsal valve from the Upper Maastrichtian sample NK5 with a loop development at the early precampagiform stage has the following dimensions: length 1.68 mm; width 1.60 mm.

Description. – The shell is large, and the outline is pointed oval to subcircular in the juvenile stages. Later in the ontogeny the outline is characteristically pentangulate. The juvenile shells are plano-convex to flat biconvex and change during ontogeny to biconvex. The maximum width is posterior to the mid-length of the shell. The anterior commissure is rectimarginate in the juveniles and later on uniplicate. The shell surface is characteristically covered with regular, closely spaced tubercles. The tubercles are cylindrical in the juvenile stages and turn conical later on. The length of the tubercles is around twice their diameter. The dorsal and ventral valves are not equally sculptured. The dorsal valve is smooth until a shell width of about 1.00 mm, where the first tubercles are formed. The ventral valve is, on the contrary, covered with tubercles throughout ontogeny. The beak is acute and erect in juveniles, and changes to slightly incurved and often attrite in the later stages. The foramen is large, trapezoidal and submesothyridid in the juveniles, and later relatively small and permesothyridid. The deltidial plates are distinct, narrow ridges whose anterior edges change into two triangular plates that limit the foramen anteriorly.

The inner socket ridges are short and relatively low and parallel. A well-developed hinge plate is present and is an anterio-posterior elongated concave plate anteriorly united with the median septum. The frontal margin of the hinge plate runs obliquely against the median septum from the inner socket ridge. At a shell width of about 1.60 mm, the brachidium is in an early precampagniform stage and consists of a low median septum and two short crura developed from the anterior margin of the inner socket ridge. Furthermore, two short antero-laterally directed branches are developed in the anterior of the median septum (Pl. 20:6). According to Owen (1970), *K. pentangulata* reaches a kingeniform stage in the adult loop development. The shell is relatively thin.

Remarks. – *Kingena pentagulata* is distinguished from other species of *Kingena* in the umbonal features (Owen 1970, p. 42). *K. pentangulata* has a dome-like umbo with a strong tendency to labiation of the foramen, and the interarea is very short. Furthermore, *K. pentangulata* is characterized by the evenly distributed tubercles on the shell surface and by the pentangular outline.

Occurrence. – *K. pentangulata* occurs in the Upper Maastrichtian samples NK1, NK3, NK4, NK5, NK6, NK7, NK8 and NK9. The species is represented in a low number of individuals, and the species becomes extinct at the Maastrichtian–Danian boundary.

Genus *Dalligas* Steinich 1968

Type species. – *Dalligas nobilis* Steinich 1968, by original designation.

Dalligas nobilis Steinich 1968
Pl. 18:6, 7

Synonymy. – ☐ 1968 *Dalligas nobilis* gen. et. sp. nov. – Steinich, pp. 336–347, Figs. 1–3, Pl. 1. ☐ 1984 *Dalligas* sp. – Surlyk & Johansen, Fig. 1.

Material. – Fifteen dorsal valves, five ventral valves and a number of fragments. The largest specimen is from the Upper Maastrichtian sample NK5 and is represented by a dorsal valve with the following dimensions: length 2.48 mm; width 3.04 mm.

Description. – The shell is small and thin with an outline changing from elongated subquadratic in the juvenile stages to broadly subquadratic in large forms. The maximum width is approximately at the mid-length of the shell, and the anterior commissure is straight. The shell is flatly biconvex with a very flat dorsal valve, and the auricles are distinct. The shell surface contains 3–4 distinct growth lines on an otherwise smooth surface. The hinge line is broad and straight, the beak is erect, and the area is narrow and clearly delimited. The foramen is large, hypothyridid, subtriangular and very attrite, laterally confined by two narrow, high deltidial plates. Anteriorly, the foramen is rounded by protruding cardinalia. The pedicle collar is well developed and constitutes about two-thirds of the height of the beak. The hinge is well developed, and the teeth are acute and triangular. The inner socket ridges are high and short. Crura, crural processes, and descending arms are lacking. The dorsal median septum is a thin, short, and pointed plate which rises around mid-length of the shell from a long low-lying base. The spicular skeleton was relatively strong and is often recrystallized; it reveals that the lophophore was probably a schizolophe.

Remarks. – The species is assigned to *Dalligas nobilis* Steinich due to the large foramen, the long distinct pedicle collar and the missing crura, crural processes and descending arms.

According to the diagnosis of *D. nobilis* (Steinich) this species has a weak rib pattern along the shell margin, but the material from Nye Kløv does not include specimens large enough to possess this sculpture. *Dalligas nobilis* is difficult to separate from juvenile forms of *Magas chitoniformis*. The latter can, however, be recognized on its large punctae and well developed brachidium.

Occurrence. – *Dalligas nobilis* occurs in the upper Maastrichtian samples NK3, NK4, NK5, and NK7, in a low number of individuals. The species has not been recorded from the Lower Danian.

Dalligas sp. Steinich 1968
Pl. 19:4, 5

Synonymy. – ☐ 1984 *Dalligas*(?) sp. – Surlyk & Johansen, Fig. 1.

Material. – Eleven complete shells, five dorsal and one ventral valve, all possibly juveniles. The largest specimen (from the Lower Danian sample NK22) is represented by a dorsal valve with a length of 1.40 mm and a width of 1.68 mm.

Description. – The shell is small, smooth and relatively thick. The outline is in the earliest stages broad, rounded triangular with maximum width at the anterior shell margin. Later, the outline is subpentangular, with the maximum width anterior to the mid-length of the shell, and the shell is only slightly longer than wide. The shell is biconvex, and the dorsal valve contains in the earliest stages a broad median depression anteriorly, which gives a sulcate front commissure. A few growth lines can be traced. The hinge line is in the earliest stages relatively broad and almost straight, and later becomes very broad and straight. The beak is long, erect to slightly incurved, and the area is triangular and narrow. The foramen is rather small, broad, triangular and hypothyridid, and limited laterally by two relatively high, plate-like deltidial plates. A pedicle collar is well developed. The hinge is somewhat feebly developed, the teeth are small and blunt, and the inner socket ridges are low, short and parallel. The hinge plates are developed as low ridges anteriorly converging from the anterior margin of the inner socket ridges. They reach the base of the median septum, where they fuse. A triangular concave thickened area is developed between the hinge plates.

The crura were probably short and slender. The median septum is thin, long and low, and is approximately half the length of the shell. The nature of the lophophore is unknown.

Remarks. – The broad triangular outline and the median depression in the dorsal valve distinguish the juvenile of this species from other juveniles. The features of the hinge plates described show affinities to the genus *Dalligas* Steinich (see also Steinich 1968b, p. 341, Fig. 3), but the material does not allow for any further statement, and the classification is therefore somewhat uncertain.

Occurrence. – The species occurs in the Lower Danian samples NK22, NK23 and NK25. This is the first recorded occurrence of the species in the Lower Danian.

Family Terebratellidae King 1850

Subfamily Magasinae Dall 1870

Type species. – *Magas pumilus* J. Sowerby 1816, by original designation.

Genus *Magas* Sowerby 1816

Magas chitoniformis (Schlottheim 1813)

Synonymy. – ☐ 1813 *Terebratulina chitoniformis* – Schlottheim, p. 133 (cit. Faujas, Pl. 26:6). ☐ 1821 *Magas pumilus* – Sowerby, J, p. 40, Pl. 119:3–5. ☐ 1894 *Magas pumilus* – Posselt, p. 48, Pl. 1:10–11. ☐ 1909 *Magas pumilus* Sowerby – Nielsen, p. 170, Pl. 2:106. ☐ 1963a *Magas chitoniformis* (Schlottheim) – Steinich, p. 608, Fig. 8. ☐ 1965 *Magas chitoniformis* (Schlottheim) – Steinich, pp. 183–193, Figs. 280–294, Pl. 19:2, Pl. 20:1a–d. ☐ 1972 *Magas chitoniformis* (Schlottheim) – Surlyk, pp. 26, 40, Figs. 5, 11, 12, Pl. 5C. ☐ 1979 *Magas chitoniformis* (Schlottheim) –Bitner & Pisera, pp. 81–82, Pl. 7:5–6. ☐ 1984

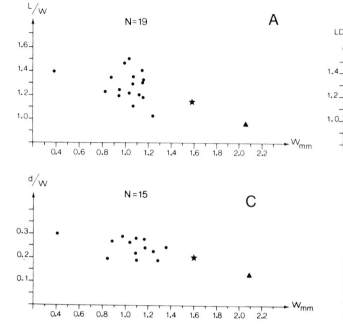

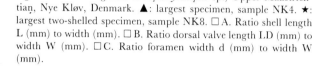

Fig. 28. Scatter diagrams of *Leptothyrellopsis* sp., Upper Maastrichtian, Nye Kløv, Denmark. ▲: largest specimen, sample NK4. ★: largest two-shelled specimen, sample NK8. □ A. Ratio shell length L (mm) to width (mm). □ B. Ratio dorsal valve length LD (mm) to width W (mm). □ C. Ratio foramen width d (mm) to width W (mm).

Magas chitoniformis (Schlottheim) – Surlyk & Johansen, Fig. 1.

For a more complete list of synonyms, see Steinich (1965, pp. 83–185).

Material. – One complete shell, eight dorsal and three ventral valves and a number of fragments. The majority of the material is represented by juveniles. It has not been possible to obtain any measurements of the largest specimen in the material.

Description. – The material does not allow for any adequate description of the ontogenic development of the species. For a more detailed description, see Steinich (1965). The material from Nye Kløv is consistent with this description.

The material shows, however, that the shell is relatively large, and the juvenile outline is elongated oval. Larger specimens are subcircular. The maximum width is around the mid-length of the shell. The auricles are indistinct. The shell is biconvex in juveniles, but during growth turns planoconvex with the ventral valve of very high convexity. The anterior commissure is straight to broadly sulcate. The shell surface is smooth, and growth lines are distinct. The hinge line is straight and relatively long, which gives the posterior margin of the dorsal valves of the juveniles characteristic shoulders. The beak is relatively low, blunt and erect but becomes distinctly incurved during growth. The area is small and triangular, plane in the juvenile stages but turns concave in the later stages of ontogeny. The foramen is subtriangular, large and hypothyridid in the juveniles and later triangular and submesothyridid. The deltidial plates are narrow, distinct ridges, and because of the incurved beak the deltidial plates are invisible from the outside in large forms. The pedicle collar is well developed, and the hinge is strongly built. The inner socket ridges are strong, high, short, and reach far behind the posterior margin of the dorsal valves. The outer socket ridges are low, and the cardinal process is either very small or missing. The brachi-

dium is described in detail by Steinich (1965). The dorsal median septum is strong and high, and the ventral valve contains a low median ridge in the anterior part of the valve floor. The shell is thin in the juvenile stages but is later very thick, especially the posterior part of the shell.

Remarks. – The juveniles of *Magas chitoniformis* resemble very much *Dalligas nobilis* Steinich. The latter, however, possess a very simple brachidium. Larger forms contain weak radial ribs, and the shell is flatly biconvex. Furthermore, the shell surface of *M. chitoniformis* contains large punctae, and the posterior margin of its dorsal valve possesses characteristic shoulders.

The species is index species for the topmost Maastrichtian *stevensis–chitoniformis* Zone (Surlyk 1970, 1982, 1984) the top of which is defined by the disappearance of *Magas chitoniformis*.

Occurrence. – *Magas chitoniformis* is found in the Upper Maastrichtian samples NK1, NK3 and NK6 in a small number of individuals. The species becomes extinct at the Maastrichtian–Danian boundary.

Family uncertain

Genus *Leptothyrellopsis* Bitner & Pisera 1979

Type species. – *Leptothyrellopsis polonicus* Bitner & Pisera 1979, by original designation.

Leptothyrellopsis sp.
Pl. 20:1–4; Figs. 28A–C, 29A–C

□ 1984 *Leptothyrellopsis* sp. – Surlyk & Johansen, Fig. 1.

Material. – Thirteen complete shells, 35 dorsal and 11 ventral valves and a number of fragments. The largest specimen is from the Upper Maastrichtian sample NK8 and has the following dimensions: length 2.24 mm; dorsal valve length 1.84 mm; width 1.60 mm; thickness 0.56 mm; and width of foramen 0.32 mm.

Description. – The species is small, smooth, and possesses an outline changing from oval to elongated oval. In juveniles the maximum width lies at the anterior shell margin, and in larger forms around the mid-length of the shell. The ratios of L/W, LD/W and d/W are shown in Fig. 28A–C. The hinge line is short and straight. The beak is suberect in the early growth stages but changes to erect and later to slightly incurved. The area is triangular and clearly limited, and the beak is large, hypothyridid, subtriangular, and laterally limited by two high narrow deltidial plates that may fuse posteriorly. The foramen is often attrite. Anteriorly the foramen is rounded by protruding dorsal cardinalia. The pedicle collar is very flat and constitutes almost the length of the beak. The hinge is well developed. The teeth are acute and triangular with deep teeth sockets and inner socket ridges which are high and short. The cardinal process is small and bulbous. The inner socket ridges are laterally concave and converge somewhat posteriorly. The basis of the anterior margins of the inner socket ridges continue into slender, ventrally converging crura. Information concerning the distal ends of the crura is lacking. The median septum rises as a short, relatively high plate. During growth it changes to a broad triangular plate that reaches behind the mid-length of the shell. In large specimens two flat wings representing the ascending branches of the brachidium occurs laterally on the posterior edge. The spicular skeleton is strong and is often recrystallized. The lophophore develops through ontogeny from a trocholophe to probably a spirolophe. The shell is punctate and relatively thin.

Remarks. – The described species is reminiscent of *Leptothyrellopsis polonicus* Bitner & Pisera in being small, smooth-shelled with an elongated outline, in possessing a large hypothyridid foramen, and in having a very long dorsal median septum. It differs, however, in possessing a pedicle collar, crura and ascending branches. The description of the genus *Leptothyrellopsis* of Bitner & Pisera (1979) from the Polish Mielnik locality of Upper Cretaceous age is, on the other hand, based on a few individuals, and the illustrated material is recrystallized and fragmented. Therefore it is uncertain how comprehensive the morphological differences between the Danish material described here and the Polish *Leptothyrellopsis polonicus* actually are. It is for the present thus preferred to refer to the Danish material to the genus *Leptothyrellopsis*. For comparison, the specimens from Nye Kløv is included on scatter diagrams of *L. polonicus* Bitner & Pisera from Mielnik (Fig. 29A–C).

　　Leptothyrellopsis sp. is similar to the juvenile form of Schloenbach's *Morrissia suessi* Bosquet (Schloenbach 1866, Pl. 39:14–15, especially Fig. 15), both forms being small and smooth- shelled with an elongated outline. Both forms also possess large hypothyridid foramens and a long median septum. A closer comparison of the interior of the two forms has not been possible.

Occurrence. – *Leptothyrellopsis* sp. occurs in the upper Maastrichtian samples NK1, NK2, NK3, NK4, NK5, NK6,

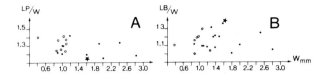
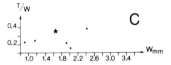

Fig. 29. Scatter diagrams of *Leptothyrellopsis* sp., Upper Maastrichtian, Nye Kløv (●; ★: largest two-shelled specimen, sample NK81), and *Leptothyrellopsis polonicus* Bitner & Pisera, Upper Campanian – Lower Maastrichtian, Mielnik, Eastern Poland (○; data from Bitner & Pisera 1979, Fig. 5. □ A. Ratio shell length LP (mm) to width W (mm). □ B. Ratio dorsal valve length LB (mm) to width W (mm). □ C. Ratio thickness T (mm) to width W (mm).

NK7, and NK8. The species occurs in a small number of individuals and has not been found in the Danian.

Changes in the brachiopod faunas at Nye Kløv

The washing residue curve

The white chalk from the Upper Maastrichtian part of Nye Kløv is relatively poor in macrofossils, containing only about 1 wt% skeletal remains larger than 0.5 mm (Håkansson & Thomsen 1979; Johansen 1982; Fig. 6 herein). Coccoliths and coccolith detritus compose the remaining part of the chalk. The pelagic chalk from the lowermost Danian is even less fossiliferous, containing considerably less than 1 wt% skeletal material, and so is the thin clay bed and the marly chalk succeeding the end-Maastrichtian event. From about 2 m above the basal Danian clay bed, the content of skeletal remains increases gradually, reaching a maximum in a bryozoan limestone around 7 m above the boundary. Here the skeletal fragments constitute up to 10 wt% of the sediment. In the upper part of the section the content of fragments drops to around 1 wt% again in the Lower Danian pelagic chalk.

The fauna in the washing residue

The Maastrichtian chalk of northwestern Europe is typically dominated by bryozoans (Håkansson *et al.* 1974; Surlyk 1972; Surlyk & Birkelund 1977). The washing residue curve, also referred to as the benthos curve, gives an approximate picture of the quantitative distribution of bryozoans and thus indirectly reflects the amount of hard substrate available for the brachiopods. Besides bryozoans and brachiopods, fragments of echinoids, bivalves, crinoids, serpulids and foraminiferans are common in the washing residue (e.g. Birkelund & Håkansson 1982). The larger potential substrates, which normally are not represented in the washing residue, include echinoid tests, bivalve shells, sponges and soft bodied organisms such as ascidians and horny worm tubes.

In the washing residue of the Maastrichtian chalk from Nye Kløv the bryozoans constitute between 50 and 80 wt% of the fauna (Fig. 6). The remaining skeletal material is dominated by fragments of echinoids.

The lowermost Danian is totally different from the Upper Maastrichtian in faunal composition. The washing residue from the basal 2 m of the Lower Danian is dominated by skeletal parts of the crinoid *Bourgueticrinus*, and bryozoans are virtually absent (Håkansson & Thomsen 1979). Apart from the crinoids, only indeterminable remains of irregular echinoids and a few species of bryozoans and foraminiferans constitute the benthic fauna. Brachiopods are totally absent in this part of the section.

From approximately 2 m above the Maastrichtian–Danian boundary the content of bryozoans gradually increases and constitutes almost 90 wt% of the fauna in the bryozoan limestone deposited higher in the section. Brachiopods reappear in the Lower Danian chalk approximately 3 m above the Maastrichtian–Danian boundary, where they constitute around 1 wt% of the skeletal remains. Hereafter, the brachiopod content increases to between 5 and 10 wt% in the most bryozoan-rich horizons. In the more benthos-rich horizons large amounts of echinoids are also present along with minor amounts of crinoids, bivalves, serpulids and foraminiferans. In the upper part of the sequence the brachiopods gradually drop in frequency again.

The benthos as a substratum for the brachiopods

The brachiopod density curve (number of individuals per kilogramme sample weight) and the benthos curve (amount of skeletal remains larger than 0.5 mm per kilogramme sample weight) are almost parallel (Fig. 6). This feature was previously observed in the Maastrichtian chalk of Rügen (Steinich 1965) and the Danish Maastrichtian chalk (Surlyk 1972).

In the Maastrichtian of Nye Kløv these two curves are parallel, whereas in the Lower Danian the brachiopod curve seems to develop subsequent to the benthos curve, most likely in consequence of the change in bottom conditions at the time of the Maastrichtian–Danian transition. The crinoid *Bourgueticrinus*, dominating the benthos in the Lower Danian, was among the first benthic organisms colonizing the soupy macrohabitat. The crinoid was adapted to a mode of life on soft bottoms using cirri at the end of the column for attachment (Rasmussen 1961). The group of bryozoans occurring in the basal Danian is also dominated by free-living species, which were similarly specialized to a mode of life on the soft chalk bottom. The dominating species, *Pavolunites* n.sp., lay unattached on the bottom supported by stiff marginal setae. The remaining species and their colonies were supported above the bottom by uncalcified rootlet structures (Håkansson & Thomsen 1979). These faunas disappear together with *Bourgueticrinus* in higher, more bryozoan-rich horizons.

Both the crinoids and the free-living bryozoans were probably too widely dispersed and provided too unstable a substratum to permit a stable brachiopod fauna to be established after the end-Cretaceous devastation of the fauna.

Rigidly erect, branching bryozoan colonies attached by a solid basis appear higher in the section and eventually dominate the fauna in the Danian bryozoan community. This type of community constitutes the typical Lower Danian bryozoan limestone. The brachiopods did not colonize the chalk environment until this stable bottom fauna of bryozoans was established, and increased hereafter rapidly in both density and diversity. The same ecological succession for the colonization of the chalk has been described by Surlyk & Birkelund (1977).

The brachiopod fauna

The Maastrichtian brachiopod fauna of the chalk of northwestern Europe was highly specialized. From homology and inferred functional morphology seven main adaptive groups are recognized (Surlyk 1972):

group Ia, minute species attached with a pedicle and able to utilize very small hard substances as substrates;

group Ib, medium to very large species attached with a pedicle and confined to large, hard substrates;

group Ic, medium-sized species attached directly to the sediment by a rooted-type pedicle;

group II, medium to large free-living species with pedically attached juvenile stages;

group III, burrowing species;

group IVa, cementing species able to utilize very small substrates and essentially free-living as adults; and

group IVb, cementing species confined to large, hard substrates.

Changes in adaptive groups

Ia. Minute, pedically attached forms. – This adaptive group overwhelmingly dominates in both stages. In the Upper Maastrichtian 15 species are present, and in the Lower Danian 19 species (Fig. 30).

Furthermore, of the five articulate species crossing the Maastrichtian–Danian boundary at Nye Kløv, four belong to group Ia: *Terebratulina longicollis* (*T.* cf. *longicollis*), *Argyrotheca hirundo*, *A. stevensis* and *Aemula inusitata*.

The species diversity is virtually the same in the brachiopod faunas from the Upper Maastrichtian and the Lower Danian (Figs. 6, 7), but the species present are almost totally different. At the generic and higher levels there is a marked difference in the dominance: Cancellothyridid brachiopods (genera related to the genus *Terebratulina*) dominate in the Maastrichtian and *Argyrotheca* species in the Danian. In the Maastrichtian of Nye Kløv the cancellothyridid group includes eight species: *Terebratulina chrysalis* (Schlottheim), *T. faujasii* (Roemer), *T. gracilis* (Schlottheim), *T. longicollis* Steinich, *Gisilina jasmundi* Steinich, *Rugia acutirostris* Steinich, *R. tenuicostata* Steinich and *Meonia semiglobularis* (Posselt). Of these *Terebratulina faujasii*, *T. longicollis*, *Gisilina jasmundi*, *Rugia acutirostris* and *R. tenuicostata* all belong to group Ia.

Next in importance is a more homogenous group of five species belonging to the genus *Argyrotheca*: *A. bronnii* (Roemer), *A. coniuncta* Steinich, *A. danica* (de Morgan), *A. hirundo* (Hagenow) and *A. stevensis* (Nielsen). These species

Adaptive groups, Nye Kløv

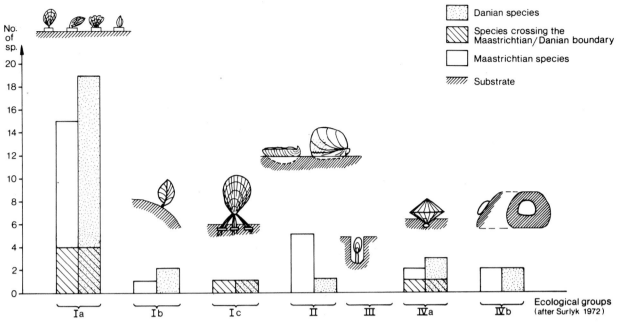

Fig. 30. Comparison of the adaptive groups of the Upper Maastrichtian and Lower Danian brachiopod faunas at Nye Kløv (from Johansen 1982). See pp. 46–50 for comments.

all belong to group Ia. A typical Upper Maastrichtian sample contains 3–6 cancellothyridid species of which 1–3 belong to the adaptive group Ia, and 1–3 *Argyrotheca* species.

In the Lower Danian of Nye Kløv the situation is reversed. Here the dominant genus is *Argyrotheca*, which is represented by at least eight species: *Argyrotheca* aff. *bronnii* (Roemer), *A. hirundo* (Hagenow), *A. stevensis* (Nielsen), *A.* aff. *stevensis* (Nielsen), *A. dorsata* (Nielsen), *A.* cf. *faxensis* (Posselt), *A. vonkoeneni* (Nielsen) and *A. armbrusti* (Schloenbach). Here the cancellothyridid group includes seven species: *Terebratulina chrysalis* (Schlottheim), *T. kloevensis* n.sp., *T.* cf. *longicollis* Steinich, *T.* aff. *rigida* (Sowerby), *Rugia flabella* n.sp., *R. latronis* n.sp. and *Rugia* sp., which all, except *T. chrysalis*, belong to group Ia.

A typical Lower Danian sample contains 1–3 cancellothyridid species and 5–8 species of *Argyrotheca*. Examination of a large number of samples from Lower Danian bryozoan chalk and bryozoan limestone, e.g. from Stevns Klint and from Maastrichtian bryozoan limestone and chalk from various northwest-European localities show the same pattern (Surlyk 1972; Johansen 1986, Johansen & Surlyk, unpublished data)

The specific composition of the brachiopod fauna is, for the Maastrichtian, rather constant for each stratigraphic zone (Surlyk 1972), but the abundance of each species varies considerably.

In the majority of the samples one to three species dominate in number of individuals (Figs. 31, 32, 33). The dominance of certain species usually reflects stratigraphical differences, if the collected samples represent a sufficiently long span of time to smoothen out a possibly patchy distribution of the brachiopods. A patchy distribution should be visible only if the collected samples cover short time spans and a large number of widely spaced samples from the same strati-

grahic levels are compared. Though adequate comparative material from contemporaneous sections of a similar sedimentology is largely lacking, it is here suggested that the observed differences in density and diversity of the species of *Argyrotheca* and the group of cancellothyridid species besides their stratigraphic implications also reflect differences in the adaptability of the two groups.

In density and diversity the cancellothyridid species and *Argyrotheca* species are the most dominant within group Ia (Figs. 7, 31, 32). There are, however, marked differences in the distribution of the number of species and number of individuals of the two groups in Nye Kløv. The cancellothyridid brachiopods are represented in nearly all samples in a low but rather constant number of species. Compared to the total density of the articulate brachiopod fauna (Fig. 31), this group is low in number of individuals in the samples of highest density (samples NK22 and NK25). On the contrary they are comparatively high in number of individuals in samples of low density (e.g. samples NK27 to NK30).

In the Upper Maastrichtian, *Terebratulina longicollis* is among the most abundant of the species present and is dominant in the samples NK3, NK6, NK7 and NK8 along with *Meonia semiglobularis*. In the Lower Danian, *Rugia flabella* is the most abundant of the cancellothyridid brachiopods and is dominant in the samples NK18, NK19, NK21, NK23 and NK24.

Both the living and fossil cancellothyridid species are characterized by relatively high beaks and relatively large foramina, well-developed hinges and lack of hinge plates. The living forms are attached to the substratum by a rather long, rigid pedicle containing rootlets of varying size, shape and placing. They seem to possess a rich variety of modes of life, varying use of the beak as a stabilizing organ and varying degree of substrate dependance (e.g., Rudwick 1965;

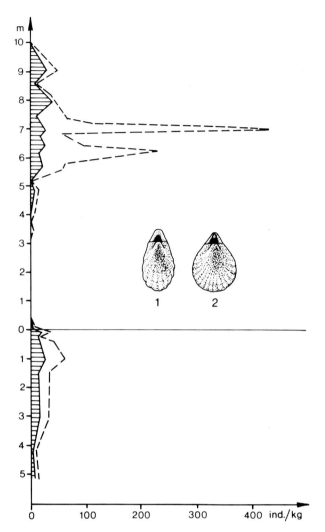

Fig. 31. Abundance of cancellothyridid brachiopods in the Nye Kløv section (hatched area). The dashed line shows the total number of brachiopod specimens per kg sample weight. In the Upper Maastrichtian *Terebratulina longicollis* Steinich (species 1) is the dominant species and in the Lower Danian *Rugia flabella* n.sp. (species 2). (From Johansen 1982.)

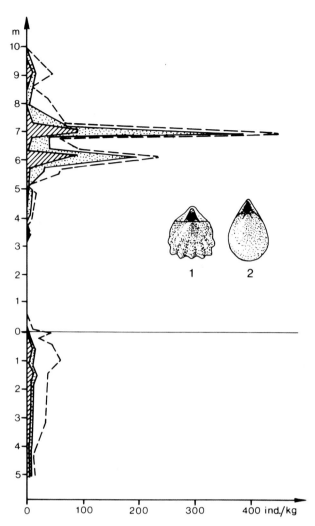

Fig. 32. Abundance of the *Argyrotheca* group (hatched area) and the group of minute, smooth-shelled, pedically attached forms (dotted area, superimposed on the hatched area). The dashed line shows the total number of brachiopod specimens per kg sample weight at Nye Kløv. In the Upper Maastrichtian neither of the groups are dominant. In the Lower Danian *Argyrotheca stevensis* (Nielsen) (species 1) is dominant in the *Argyrotheca* group. *Cryptopora perula* n.sp. (species 2) is dominant in the group of smooth-shelled forms. (From Johansen 1982.)

Surlyk 1972; Richardson 1981a, b; Curry 1982; Cooper 1973a, b, 1979, 1981a, b).

The cancellothyridid brachiopods occur throughout the Maastrichtian as well as in older chalk from the Upper Cretaceous of northwestern Europe. They seem, however, to be most abundant in chalk deposited in deeper-water, offshore environments (Johansen 1986). The relatively wide substrate tolerance of this group thus probably partly explains its survival capacity.

The abundance of the *Argyrotheca* group in Nye Kløv, in contrast to that of the cancellothyridid group, follows the benthos curve to a very large extent (Fig. 32). In the Upper Maastrichtian the group of *Argyrotheca* constitutes a very subordinate part of the brachiopod fauna. Two abundance maxima are seen in the Lower Danian samples NK22 and NK25. These maxima coincide with a prominent peak in bryozoan density (samples NK21 to NK27), and both maximae are dominated by specimens of *Argyrotheca stevensis* and *A. hirundo*.

Both living and fossil species of *Argyrotheca* are characterized by their very large foramen, withdrawn beak, broad straight hinge line, and wide area. The living forms are pressed tightly against the substratum by means of a very short, stout and muscular pedicle, the commissure of the shell thus placed perpendicular to the substrate.

Living species of *Argyrotheca* have a preference towards attachment to certain species of, e.g., bryozoans, or towards certain types of microhabitats. This was probably also true for the fossil species (Ulla Asgaard, personal communication, 1983). Among the rich variety of substrates used by Recent species of *Argyrotheca* are the ahermatypic corals *Lophelia* and *Dendrophyllia*, solitary coral thecae and crustaceans (Atkins 1960), scleractinians and demosponges (Asgaard & Stentoft 1984), algae (Rudwick 1962), bryozoans and shell fragments.

There is a dominance of species of *Argyrotheca* in Recent well-aereated, shallow-water environments. Recent species are normally not found below 200 m depth (e.g. Cooper

1973a, b, 1979; Logan 1975, 1979; Thomson 1927; Asgaard & Stentoft 1984). Fossil species of *Argyrotheca* most likely have a similar mode of life (e.g. Surlyk 1972). This group of brachiopods very commonly occurs in the offshore sublittoral sediments of the Danian of Denmark and in the nearshore Maastrichtian–Danian sediments of Holland and Belgium, the Danian sediments characteristically always more rich in *Argyrotheca* species than comparable Maastrichtian sediments (Bosquet 1859; Asgaard 1968; Johansen, unpublished data). In the deeper-water offshore chalk from the Maastrichtian of northwestern Europe, species of *Argyrotheca* are on the other hand much less common (Johansen 1986).

The rich *Argyrotheca* fauna present in the Lower Danian chalk of Nye Kløv may thus have been developed from forms that could survive in shallow waters on other types of substrates than the chalk.

In addition to the cancellothyridid brachiopods and species of *Argyrotheca*, a morphologically and taxonomically heterogenous group of small, smooth-shelled, and pedically attached species occurs throughout the Upper Maastrichtian and Lower Danian of Nye Kløv (Fig. 32). In the Upper Maastrichtian this group consists of *Scumulus inopinatus* Steinich, *Scumulus*(?) sp., *Aemula inusitata* Steinich, *Dalligas nobilis* Steinich and *Leptothyrellopsis* sp. In the Lower Danian *Platidia* sp., *Aemula inusitata*, *Dalligas* sp., *Cryptopora perula* n.sp. and *Gwyniella persica* n.sp. are present.

The very irregular shell, the large amphithyridid foramen and the internal structure of *Aemula inusitata* and *Platidia* sp. suggest that these had a mode of life comparable to that of Recent species of *Platidia* and *Amphithyris*, which press their dorsal valves firmly against the substrate (Thomson 1927; Atkins 1959; Surlyk 1974).

The great variability of the amphithyridid foramen of *Scumulus inopinatus* and the large, often attrite foramen of *Leptothyrellopsis* sp. also suggest a similar mode of life to that of *Platidia* sp.

Living species of *Cryptopora*, e.g. *Cryptopora gnomon* (Jeffreys), are considered to have rested on a soft bottom surface in a more or less posterior-downwards position tethered upstream by a long, slender pedicle with a few short distal rootlets (Curry 1983). As *Cryptopora perula* from the Lower Danian in its morphology is close to its Recent relatives, it may have had a similar mode of life, although this is difficult to prove.

In the Upper Maastrichtian of Nye Kløv this heterogenous group is common and constant, individuals of *Scumulus inopinatus* being the most abundant. In the Lower Danian *Gwyniella persica* and *Cryptopora perula* are most frequently occurring in the most benthos-rich horizons (Fig. 32).

Ib. Larger forms attached with a pedicle to hard substrates. – Both *Neoliothyrina* and *Kingena* are medium to large species provided with pedicle opening suitable for a normal functional pedicle. The surroundings of the foramen are furthermore often strongly worn, indicating that the pedicle was short and that the shell was pressed tightly against the substrate. Whereas *Kingena pentangulata* (Woodward) is fairly common, species of *Neoliothyrina* have not been recorded from the Upper Maastrichtian of Nye Kløv (Fig. 30). Surlyk (1972) mentions that *Neoliothyrina* is very rare in the Danish chalk and that this is probably due to the rarity of suitable large

hard substrates. This is most likely also true for Nye Kløv. In the Lower Danian, however, a species referred to as *Neoliothyrina*? is characterized by its large, thin shell and its strongly worn beak, thus being a typical member of group Ib. The Lower Danian species of *Carneithyris* are, according to Asgaard (1975), inferred to have been attached to the substrate by a functional pedicle and consequently belonging to group Ib, as opposed to the Upper Maastrichtian *Carneithyris subcardinalis* which during ontogeny closed its pedicle opening by secondary shell.

Ic. Medium-sized species with pedicle rooted in sediment. – *Terebratulina chrysalis* (Schlottheim), the most common brachiopod in the Danish chalk occurring in a rich variety of depositional environments, is present in the Upper Maastrichtian as well as in the Lower Danian of Nye Kløv (Fig. 30). The species is among the first brachiopods to recolonize the Lower Danian chalk sea bottom and is also among the last to disappear. Surlyk (1972) suggest that the pedicle of this species was distally split into fine rootlets, thus being capable of rooting itself directly in the fine coccolith muds. This suggestion is supported by the fact that certain Recent relatives in soft substrates exhibit a similar mode of life (Rudwick 1961; Cooper 1973a, b, c; Stewart 1981; Curry 1982).

II. Larger free-living species with attached juvenile stages. – At least five species belonging to this group are present in the Upper Maastrichtian of Nye Kløv: *Cretirhynchia* sp. [*C. retracta* (Roemer) and/or *C. limbata* (Schlottheim)], *Carneithyris subcardinalis* (Sahni), *Terebratulina gracilis* (Schlottheim), *Magas chitoniformis* (Schlottheim) and *Meonia semiglobularis* (Posselt) (Fig. 30). For the Danish Maastrichtian chalk as such, ten species from this group are recorded. These are, besides the mentioned species, one or two additional species of *Cretirhynchia*, *Gemmarcula humboldtii* (Hagenow), *Thecidea papillata* (Schlottheim) and *Trigonosemus pulchellus* (Nilsson). Of these, *Trigonosemus pulchellus* is stratigraphically restricted to the Upper Lower Maastrichtian (Surlyk 1972, 1982; Johansen 1986), and several of the species are moreover restricted to more shallow-water facies. In the Lower Danian of Nye Kløv, on the other hand, only one indetermined species of *Cretirhynchia* is present from this group.

Group II comprises species which as adults lie unattached on the chalk sea floor. At the early growth stages these brachiopods possessed the same general terebratulinid shape as the brachiopods from group Ia, i.e. a shell that is biconvex, longer than wide and provided with a normal functional foramen. During ontogeny, however, the pedicle opening of this group was either narrowed into a 'pin-hole' foramen (e.g. *Terebratulina gracilis*, *Magas chitoniformis*, and *Meonia semiglobularis*), closed by secondary shell or at least hidden by the incurved umbo (e.g. species of *Cretirhynchia*, *Carneithyris subcardinalis*, *Trigonosemus pulchellus* and *Gemmarcula humboldtii*).

It is characteristic that two different soft-stratum adaptations have developed within group II. The shells of the rhynchonellid species are laterally expanded in what often is termed a 'snowshoe' morphology, and the remaining species attain as adults a gryphaete or hemisphaerical reclining shape characteristic of what is termed an 'iceberg' morphology (Rhoads 1970; Thayer 1975a; Carter 1972; Stenzel

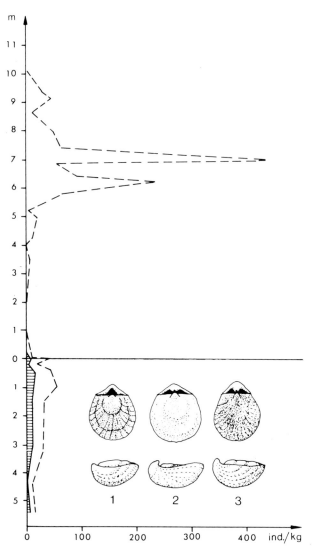

Fig. 33. Abundance of three species of the secondarily free-living hemispherical reclining brachiopods in the Nye Kløv section. The dashed line shows the variation in the total number of brachiopod specimens per kg sample weight. Species 1: *Meonia semiglobularis* Posselt; species 2: *Magas chitoniformis* (Schlottheim); species 3: *Terebratulina gracilis* (Schlottheim). (From Johansen 1982.)

1971; Jablonski & Bottjer 1983).

Carneithyris subcardinalis is the only exception to these two morphologies in its strongly biconvex to almost sphaerical shell. The cardinalia of this species are secondarily thickened, the centre of gravity thus positioned posteriorly. The species is the only common large brachiopod in the chalk and seems morphologically to fit the ideal shape of a free-living brachiopod, e.g. a self-righting tumbler (Surlyk 1972).

It is noteworthy that only one species (*Cretirhynchia* sp.) from group II is present in the Lower Danian chalk of Nye Kløv, although the sediment here texturally is equivalent to the Upper Maastrichtian chalk.

Terebratulina gracilis, Meonia semiglobularis and *Magas chitoniformis* are in density a dominant group in Nye Kløv (Fig. 33). *Meonia semiglobularis* is by far the most abundant and is dominant in the samples NK3, NK6, NK7, NK8 and NK9 together with *Terebratulina longicollis*. The Lower Danian chalk from the localities Kjølbygård, Eerslev, Bjerse, and Dania has been analysed for brachiopods, too, but apart

from undeterminable rhynchonellid fragments, this group is not represented.

III. Burrowing forms. – In the Danish Maastrichtian chalk only one small, thin-shelled species of *Lingula* is recorded (Surlyk 1972), *Lingula cretacea* (Lundgreen). The genus is represented neither in the Upper Maastrichtian or Lower Danian samples from Nye Kløv, either due to the fragility of its shell (it is only rarely found in the washed samples) or owing to its palaeoecological restrictions to sediments of a more shallow-water nature (Paine 1970; Thayer & Steele-Petrovic 1975).

IV. Cemented forms. – Both group IVa (species cemented to the substratum by a minute attachment surface and essentially free-living as adults) and IVb (species cemented to and confined to large, hard substrates) comprise inarticulate brachiopod species, the systematics of which are not dealt with in detail in the present paper.

Group IVa is in the Upper Maastrichtian of Nye Kløv represented by two species of *Isocrania, Isocrania costata* (Sowerby), and *Isocrania* aff. *costata* 1, both of which have very small attachment surfaces. *Isocrania costata* crosses the Maastrichtian–Danian boundary and occurs in the Lower Danian together with *Isocrania* aff. *costata* 2 and *Isocrania* sp.

Group IVb in the Upper Maastrichtian consists of *Crania* aff. *craniolaris* (Linnaeus) and *Craniscus* sp., and in the Lower Danian of *Crania tuberculata* (Nielsen), and *Crania* sp. From the Danish Maastrichtian chalk as such at least seven species of this group are recorded (Surlyk 1972). The sparsity in Nye Kløv is partly due to their attachment to large, hard substrates which normally are not found in the washing residues, and partly owing to the fact that these forms palaeoecologically are also numerous in more shallow-water facies.

Brachiopod extinctions across the Maastrichtian–Danian boundary

A range chart of all the brachiopod species found in Nye Kløv is shown in Fig. 7. The chart is an updated version of the scheme presented by Johansen (1982) and Surlyk & Johansen (1984). It must be emphasized that the range chart presents the raw data from Nye Kløv only, this being the most complete of the Maastrichtian–Danian boundary localities in higher latitudes. The range chart hence does not mirror a regional range pattern of the individual species. All of the Maastrichtian species shown on the chart have thus their first appearances much earlier in the chalk (Surlyk 1982; Johansen 1986) and some of the Lower Danian species recorded from Nye Kløv occur likewise in younger Danian strata (Asgaard 1968; Johansen, unpublished data). The Maastrichtian part of the sequence contains 27 species, all of which are well known from the neighbouring localities, Kjølbygård and Bjerre (Surlyk 1969, 1982, unpublished data). Furthermore, several additional species are known from the highest Maastrichtian in the more benthos-rich chalk at Stevns Klint (Surlyk 1969, 1972). The Danian part of the sequence contains 42 species that can be subdivided into three stratigraphical groups.

(1) Thirteen species, all of which are known from the Upper Maastrichtian, are restricted to the basal 3 cm thick Danian clay bed. These are: rhynchonellid sp., *Terebratulina faujasii*, *T. gracilis*, *Meonia semiglobularis*, *Rugia acutirostris*, *Argyrotheca bronnii*, *A. coniuncta*, *Scumulus inopinatus*, *Kingena pentangulata*, *Leptothyrellopsis* sp., *Isocrania* aff. *costata* 1, and *Craniscus* sp., and possibly also *Gisilina jasmundi*. They have never been found higher in the Danian sequence and are furthermore broken, whitish in colour and worn. They are almost certainly contained in reworked pebbles of Maastrichtian chalk, which commonly occur in the basal Danian clay bed. These 13 species are treated as belonging to the Maastrichtian fauna and are not included in the Danian.

(2) Six of the species are common to the Maastrichtian and the Danian, and represent forms that have crossed the boundary. They are: *Terebratulina chrysalis*, *Isocrania costata*, *Argyrotheca stevensis*, *A. hirundo*, *Terebratulina* cf. *longicollis* and *Aemula inusitata*. The specific assignment of *T.* cf. *longicollis* is uncertain. It may well represent a new Danian species.

(3) The remaining 23 species appear for the first time in the Danian.

The true indigenous Early Danian brachiopod fauna of Nye Kløv is thus represented by about 29 species. A maximum of six of these species occur also in the Upper Maastrichtian.

At the species level, there is an almost complete turn-over of the brachiopod fauna at the Maastrichtian–Danian boundary, where at least 75% of the Maastrichtian species become extinct. At the generic and higher levels the replacement is less prominent. In the Upper Maastrichtian of Nye Kløv 16 genera are present, and of these at least seven disappear at the top of the Maastrichtian. The Lower Danian contains 13 genera, two of which are new (Johansen 1982). This agrees well with Naidin (1979), who also concluded that the greatest taxonomical changes across the Maastrichtian–Danian boundary for the brachiopods as well as for other groups occurs at low taxonomical levels.

For the end-Cretaceous brachiopod fauna of Nye Kløv three features are important: (1) The disappearance is sudden and coincides with the Maastrichtian–Danian boundary even on a millimetre scale for most species (Figs. 6, 7), (2) it is significant that there are no early warning signals in the form of a gradual decrease in species diversity, or change in population structure at the end of the Cretaceous, and (3) the most specialized species, in particular the secondarily free-living, hemispherical forms, become extinct at the boundary.

The diversity range of the Upper Maastrichtian brachiopod fauna from Nye Kløv (on the average 15 species per sample) is fully comparable to diversities of other Maastrichtian faunas from corresponding lithologies (Surlyk 1972; Steinich 1965; Johansen 1986; Johansen & Surlyk, unpublished), and does not seem to be affected towards the boundary (Fig. 6).

Surlyk & Johansen (1984) carried out size-frequency studies of the two most abundant species of the Upper Maastrichtian of Nye Kløv, *Meonia semiglobularis* and *Terebratulina longicollis*, in order to test possible changes in the population structures of the brachiopod fauna in the top of the Maastrichtian. The factors of population dynamics are the most important in shaping the size-frequency of the fossil brachiopod populations, and differences in size-frequency distributions are often due to differences in recruitment mode of the individual species (McCammon 1970; Richards & Bambach 1975; Noble & Logan 1981). The effect of minor fluctuations

Terebratulina longicollis

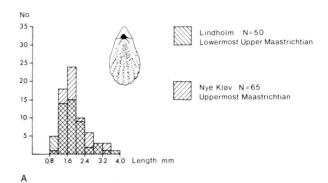

A

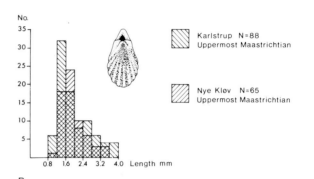

B

Fig. 34. Size-frequency distribution of *Terebratulina longicollis* Steinich from Nye Kløv superimposed on the corresponding distributions of two populations from the lowermost Upper Maastrichtian (Lindholm) and the uppermost Maastrichtian (Karlstrup) respectively. All distributions are characterized by high juvenile mortality. (From Surlyk & Johansen 1984.)

eventually leading to a major crisis would be expected to appear in the population structures of the individual species. The two species from Nye Kløv were compared with published size-frequency histograms for the same species from Denmark (Fig. 34). The histograms for the individual samples are remarkably uniform, and both species show size-frequency distributions that are strikingly similar to the distributions from other Maastrichtian samples studied. The same pattern is apparent in all other localities in Denmark that have been studied (Surlyk 1969, 1972, 1979, 1982, unpublished data).

Regardless of their taxonomic affinities, almost all the benthic inhabitants of the Upper Cretaceous muddy bottoms exhibit morphological adaptations for soft substrata. This is the case for the bivalves and the gastropods, as well as for the brachiopods (Rhoads 1970, 1974; Carter 1968, 1972; Stanley 1970; Surlyk 1972, 1973, 1974; Thayer 1975a; Gould 1977; Skelton 1979; Jablonski & Lutz 1980; Bottjer 1981; Jablonski & Bottjer 1983; Sohl & Koch 1982).

Among these adaptations are: (1) Small adult size, either because high surface-to-volume ratio and thin shells enhance floatation on soft bottoms, or as a by-product of short generation times in response to periodically favourable bottom conditions or as a selection towards settlement on small hard substrates; (2) larval settlement on relatively large hard substrates securing a position above the instable sediment–water interface; (3) larval settlement on relatively small hard substrates with secondarily free-living adults assuming a broad flattened shell shape and/or readiating skeletal projections in order to distribute the weight of the organism over a large surface area ('snowshoe morphology'); and (4) larval settlement on small hard substrates with secondarily free-living adults expanded or inflated ventrally, so that the adult organism is supported by the denser sediment at depth while still retaining its sediment–water interface contact ('iceberg' or hemisphaerical reclining morphology).

It is clear that changes in the biotic and physical environment accross the Late Cretaceous–Early Tertiary were at least partially responsible for the decline of the soft-bottom assemblages in the end-Cretaceous. Increasing degree of bioturbation (Thayer 1979; Brenchley 1981) and number of predators (Vermeij 1977, 1978; Taylor 1981; LaBarbera 1981) would certainly have caused large-scale faunal replàcements. Such replacements might, however, well be effected by progressive declines in rates of speciation or increase in rate of extinction, rather than by the abrupt shifts seen in, e.g., the brachiopods (Stanley 1979; Surlyk & Johansen 1981, 1984; Johansen & Surlyk 1983).

For the brachiopods of the Northwest European Upper Cretaceous chalk, adaptation took place through the more than 30 million years that the chalk existed as a macrohabitat. The fauna reached an absolute optimum in diversity as well as in degree of specialization in the Lower Maastrichtian. This was followed by a long period of equilibrium till the end of the Cretaceous. The 'snowshoe' morphology appeared within a closely related group of rhynchonellid brachiopods at least as early as in the Coniacian. The hemisphaerical reclining adaptation, which is not among Mesozoic brachiopods, is in the chalk developed in a group of species that are not closely related and do not appear until the Upper Campanian (Johansen 1986). It is remarkable that it is the hemisphaerical recliners that became extinct at the Maastrichtian–Danian boundary and that the niche represented by these species was not exploited in the Lower Danian.

In a few regions, living brachiopods apparently still colonize soft bottoms (Richardson 1981a, b, c; Curry 1982), but articulate brachiopods living at shelf depths today are all pedunculate and restricted almost exclusively to hard substrates (e.g. Rudwick 1970; Jackson *et al.* 1971; Thayer 1975b).

Although living brachiopods can function in almost any marine environment, their capacity of adjusting to soft substrates by elimination of their functional pedicle seems to have vanished. In emphasizing the surprisingly broad range of substratum tolerances of New Zealand brachiopods, Richardson (1981a, 1986) mentions that the living species apparently all begin their benthic life attached to a hard object. Thayer (1975b) and LaBarbera (1981) have argued that all Recent brachiopods usually are dependant on lumps of coarse substrate for recruitment of soft bottoms.

Specialists can be identified anatomically by various adaptations of the valves, pedicles and muscles that make movement on a particular substrate more efficient but would be incompatible with life on any other type of substrate. Generalists are thus defined as species that can maintain a stable feeding position on substrates of any type (sensu Richardson 1986).

Loss of the pedicle as a functional organ of attachment has the result that adjustment of the position of the adult brachiopod is no longer possible. The loss of the pedicle probably also limits the possibility of further evolution (Valentine 1974). The very specialized chalk forms may thus be more susceptible to extinction during unfavourable conditions than more conservative, less specialized forms such as, e.g., *Terebratulina chrysalis*.

The species crossing the Maastrichtian–Danian boundary (*Isocrania costata*, *Terebratulina chrysalis*, *Terebratulina longicollis*, *Argyrotheca hirundo*, *Argyrotheca stevensis* and *Aemula inusitata*) were all able to use very small objects as substrates. They represent relatively featureless forms which seem to be nonspecialized and occur in large numbers as a basic stock throughout the Maastrichtian as well as in older chalk from Northwestern Europe. Moreover, they are all – except *Argyrotheca stevensis* and *Isocrania costata*, which have their first occurrences in the upper Upper Maastrichtian – very long-ranging species. *Terebratulina chrysalis* and *Aemula inusitata* thus have their first appearances in the Middle Coniacian or even earlier, *Argyrotheca hirundo* in the Lower Campanian, and *Terebratulina longicollis* in the Upper Campanian (Johansen 1986).

This group of brachiopods formed the basis for an exploitation of the Lower Danian niches. The increase in numbers of minute pedically attached forms in the Danian suggest either an increasing niche subdivision or a larger number of available microhabitats. As the Danian includes more diverse sediment types, including layers richer in bryozoans than the more uniform Maastrichtian chalk, there is some evidence in support of the latter suggestion.

The survival capacity of this basic stock becomes even more remarkable in the light of the fact that very close relatives of *Terebratulina*, *Aemula* and *Argyrotheca* seem to form the basic stock for the Recent stock of micromorphic brachiopods, too.

Nye Kløv is generally accepted as being the best preserved pelagic calcareous section representing the Cretaceous–Tertiary boundary in higher latitudes. Both the lithology and the micro- and macrofossil record compares well with the standard sequence established in El Kef, Tunesia, and in Caravaca, Spain (Romein 1977; Smit & Hertogen 1980; Smit 1982; Smit & Romein 1985).

The extinction pattern for the brachiopods is compatible with the results obtained from quantitative analysis of nannoplankton (Tappan & Loeblich 1971; Thierstein 1981; Romein 1977) and planktic foraminiferal assemblages (Smit 1982). The planktic foraminiferans and the brachiopods show similar extinction patterns (Lutherbacher & Premoli Silva 1966; Thierstein 1982; Smit 1982; Bang 1979), whereas the nannoplankton seem to have undergone final extinction later than the planktic foraminiferans (Smit & Romein 1985). Precise data are lacking for virtually all other benthic invertebrate groups, partly because of the scarcity of complete stratigraphic sections available and the limited stratigraphic resolution in most available sequences, and partly because the organisms are generally sparse and often poorly preserved.

The synchroneity of the extinctions of the plankton, the brachiopods and the bryozoans thus suggests a common cause of the extinction event.

The discovery of an apparently world-wide iridium anomaly at the Cretaceous–Tertiary boundary led Alvarez *et al.* (1980) to introduce the so-called impact theory, in which a major impact at this boundary caused environmental and biological stresses to a degree that was responsible for the mass extinctions previously reported on the boundary. This paper and a vast number of subsequent papers led to a shift in emphasis from the traditional biotic to the geochemical and geophysical aspects of the extinctions in the end-Cretaceous. The substance of these papers are very well summarized by, e.g., Alvarez *et al.* (1982, 1984) and Smit & Romein (1985).

It must be emphasized that the present paper does not intend to throw light on the initial cause of the Cretaceous–Tertiary boundary event, but it is an important contribution to the scarce biotic data so far presented on the boundary. The brachiopod fauna was in the end-Cretaceous hit by a very sudden ecological crisis, which is in agreement with the impact or any other theoretical catastrophe theory. The mass extinction of the coccoliths and pelagic foraminiferans at the end-Maastrichtian led to a total cessation of the chalk productivity (cf. Bramlette & Martini 1964). The combined effects of productivity cessation, anoxia, and beginning clay deposition, caused an almost instantaneous destruction of the chalk macrohabitat. The immediate effect of this was extinction of faunal groups that were specialized to the chalk substrate, such as the brachiopods. When chalk deposition resumed in the Lower Danian from remains of a whole new plankton flora and fauna, the habitat was rapidly restored by adaptive radiation within surviving groups. Species of these groups include forms with a wide substrate tolerance, that could survive in well-aerated shallow marine waters on other types of substrates (Surlyk & Johansen 1984).

The brachiopod extinctions thus only reflect a chain of causality, and they shed little light on the nature of the event as such.

Summary

The micromorphic articulate brachiopod fauna from the chalk of the Maastrichtian–Danian boundary section at Nye Kløv, Denmark, is investigated in terms of taxonomy, abundance, diversity and stratigraphical distribution. The main emphasis is given to the Lower Danian forms. The transition takes place in an otherwise monotonous horizontally bedded sequence of pelagic chalk.

The uppermost Maastrichtian of Nye Kløv contains 27 species, while 35 species occur in the Lower Danian. Five of the Lower Danian species are restricted to the basal Danian clay bed and are probably reworked from the Upper Maastrichtian. They are thus considered as belonging to the Upper Maastrichtian fauna.

Six species are common to the Maastrichtian and the Danian and may thus represent forms that have crossed the boundary. The remaining 23 species appears for the first time in the Danian. Of these, five species are here described as new: *Cryptopora perula* n.sp., *Terebratulina kloevensis* n.sp., *Rugia flabella* n.sp., *Rugia latronis* n.sp. and *Gwyniella persica* n.gen. et n.sp. The true indigenous Early Danian brachiopod fauna is thus represented by about 29 species.

At the species level, there is an almost total turnover of the brachiopod fauna from the Maastrichtian to the Danian. The taxonomic dominance at the generic or higher level is markedly different from the Maastrichtian to the Danian Stage. The Maastrichtian fauna is dominated by cancellothyridid brachiopods (species related to the genus *Terebratulina*). At Nye Kløv this group includes eight species of the genera *Terebratulina*, *Gisilina*, *Rugia* and *Meonia*. Next in importance is a more homogenous group of five species belonging to the genus *Argyrotheca*. In the Danian the situation is reversed. Here the dominant genus is *Argyrotheca* (eight species) whereas the cancellothyridid group includes seven species of the genera *Terebratulina* and *Rugia*.

Of the main adaptive groups recognizable in the highly specialized Maastrichtian brachiopod fauna, the group of minute pedically attached species able to utilize very small hard substrates overwhelmingly dominates in both the Maastrichtian and the Danian. In the Maastrichtian, a group of medium to large free-living hemispherical reclining species with pedically attached juvenile stages, was also a very important component of the fauna. This group became extinct at the boundary and the niche represented by these species was not exploited in the Early Danian. The species crossing the boundary were all able to use very small particles as substrates; furthermore, they represent relatively featureless, seemingly non-specialized forms that occur in large numbers throughout the Upper Cretaceous as a basic stock. The increase in number of pedically attached minute forms in the Danian suggests either an increasing niche subdivision or a larger number of available microhabitats. As the Danian includes more diverse sediment types than the uniform Maastrichtian chalk, particularly layers richer in bryozoans, there is some evidence in support of the latter suggestion.

Both the brachiopod fauna and the bryozoan fauna were adapted to a life on a chalky bottom composed primarily of coccoliths and foraminifers. A total cessation in chalk production then ultimately led to an almost instantaneous destruction of a unique macrohabitat. The immediate effect of this was the extinction of faunal groups, such as the brachiopods, that were specialized for and restricted to the chalk substrate. When chalk deposition eventually resumed in the Lower Danian, now composed of remains of a whole new flora and fauna, adaptive radiation within surviving groups led to a rapid restoration of the chalk macrohabitat. The data presented in this paper demonstrate an abrupt extinction event at the Cretaceous–Tertiary boundary and are thus in agreement with the so-called impact theory (Alvarez *et al.* 1980 and subsequent papers) or any other catastrophy theory of the end-Cretaceous extinctions. The brachiopod extinctions, however, merely represent a chain of causality and do not themselves shed light on the nature of the event as such.

References

Alvarez, L.W., Alvarez, W., Asaro, F. & Michel, H.V. 1980: Extraterrestrial causes for the Cretaceous–Tertiary extinctions. *Science* **208**, 1095–1108.

Alvarez, W., Alvarez, L.W., Asaro, F. & Michel, H.V. 1982: Current status of the impact theory for the terminal Cretaceous extinction. *Geological Society of America Special Papers 190*, 303–315.

Alvarez, W., Kauffman, E.G., Surlyk, F., Alvarez, L.W., Asaro, F. & Michel, H.V. 1984: Impact theory of mass extinctions and the invertebrate fossil record. *Science 223*, 1135–1141.

Asgaard, U. 1968: Brachiopod palaeoecology in Middle Danian limestones at Fakse, Denmark. *Lethaia 1*, 103–121.

Asgaard, U. 1970: The syntypes of *Carneithyris incisa* (Buch 1835). *Bulletin of the Geological Society of Denmark 19:4*, 361–367.

Asgaard, U. 1972: Observations on *Neolithyrina fittoni*, a rare Maastrichtian terebratulid from NW-Europe. *Bulletin of the Geological Society of Denmark 21*, 337–345.

Asgaard, U. 1975: A revision of Sahni's types of the brachiopod subfamily Carneithyridinae. *Bulletin of the British Museum (Natural History) Geology, 25:5*, 320–362.

Asgaard, U. & Stentoft, N. 1984: Recent micromorphic brachiopods from Barbados: Paleoecological and evolutionary implications. *Geobios, Memoirs special no. 8*, 29–33.

Atkins, D. 1959: The growth stages of the lophophore of the brachiopods *Platidia davidsoni* (Eudes Deslongchamps) and *P. anomioides* (Philippi), with notes on the feeding mechanism. *Journal of the Marine Biological Association of the United Kingdom 38*, 103–132.

Atkins, D. 1960: The ciliary feeding mechanism of the Megathyrididae (Brachiopoda) and the growth stages of the lophophore. *Journal of the Marine Biological Association of the United Kingdom 39*, 459–479.

Bang, I. 1979: Foraminifera in the lowermost Danian of Denmark. *In* Birkelund, T. & Bromley, R. (eds.), *Symposium on Cretaceous–Tertiary boundary events 1*, 108–114. Copenhagen.

Beck, C. 1828: Die neuesten Untersuchungen über die Kreide-Formazion der Insel Möen angestellt usw. *Zeitschrift für Mineralogie 1*, 580–582.

Beecher, C.E. 1893: Revision of the families of Loop-bearing brachiopoda. *Transactions of Connecticut Academy of Arts & Sciences 9*, 376–399.

Birkelund, T. & Bromley, R.G. (eds.) 1979: *Symposium on Cretaceous–Tertiary boundary events 1. The Maastrichtian and Danian of Denmark*, 1–210. Copenhagen.

Birkelund, T. & Håkansson, E. 1982: The terminal Cretaceous extinctions in Boreal shelf seas – A multicausal event. *Geological Society of America Special Papers 190*, 373–384.

Bitner, M.A. & Pisera, A. 1979: Brachiopods from the Upper Cretaceous chalk of Mielnik (Eastern Poland). *Acta Geologica Polonica 29:1*, 67–88.

Bosquet, J. 1859: Monographie des Brachiopodes fossiles du terrain Cretacé supérieur des Duchée de Limbourg. I-e partie. *Mémoirs géologiques de la Neerlande 3*, 1–50.

Bottjer, D.J. 1981: Structure of Upper Cretaceous chalk benthic communities, southwestern Kansas. *Paleogeography, Paleoclimatology and Paleoecology 34*, 225–256.

Bramlette, M.N. & Martini, E. 1964: The great change in calcareous nannoplankton between Maastrichtian and Danian. *Micropaleontology 10*, 291–322.

Brenchley, G.A. 1981: Disturbance and community structure: An experimental study of bioturbation in marine soft-bottom environments. *Journal of Marine Research 39*, 767–790.

Bromley, R.G. 1975: Trace fossils at omission surfaces. *In* Frey (ed.): *The study of trace fossils*, 397–428. Springer-Verlag, New York.

Bromley, R.G. 1979: Chalk and bryozoan limestone, sediments and depositional environments. *In* Birkelund, T. & Bromley, R.G. (eds.), *Symposium on Cretaceous–Tertiary boundary events 1*, 16–32. Copenhagen.

Bromley, R.G. & Gale, A.S. 1982: Lithostratigraphy of the English Chalk Rock. *Cretaceous Research 3*, 273–306.

Carter, R.M. 1968: Functional studies on the Cretaceous oyster *Arctostrea*. *Palaeontology 11*, 458–485.

Carter, R.M. 1972: Adaptions of British Chalk Bivalvia. *Journal of Paleontology 46*, 325–340.

Cheetham, A.H. 1971: Functional morphology and biofacies distribution of cheilostome Bryozoa in the Danian Stage (Paleocene) of southern Scandinavia. *Smithsonian Contributions to Paleobiology 17*, 1–87.

Christensen, L., Fregerslev, S., Simonsen, A. & Thiede, J. 1973: Sedimentology and depositional environment of Lower Danian Fish Clay from Stevns Klint, Denmark. *Bulletin of the Geological Society of Denmark 22*, 192–212.

Christensen, W.K. & Birkelund, T. 1979 (eds.): *Symposium on Cretaceous–Tertiary boundary events 2. Proceedings*. 249 pp. Copenhagen.

Cooper, G.A. 1959: Genera of Tertiary and Recent rhynchonelloid brachiopods. *Smithsonian Miscellaneous Collections 139:5*, 1–90.

Cooper, G.A. 1971: Eocene brachiopods from Eua, Tonga. *U.S. Geological Survey Professional Papers 640-F*, 1–9.

Cooper, G.A. 1973a: Fossil and Recent Cancellothyridacea (Brachiopoda). *Science Reports of the Tohoku University (Sendai, Japan). Series 2 (Geology) Spec. Vol. 6 (Hatai Mem. Vol.)*, 271–390.

Cooper, G.A. 1973b: New Brachiopoda from the Indian Ocean. *Smithsonian Contributions to Paleobiology 16*, 1–43.

Cooper, G.A. 1973c: Vema's Brachiopoda (Recent). *Smithsonian Contributions to Paleobiology 17*, 1–51.

Cooper, G.A. 1979: Tertiary and Cretaceous brachiopods from Cuba and the Caribbean. *Smithsonian Contributions to Paleobiology 37*, 1–30.

Cooper, G.A. 1981a: Brachiopoda from the southern Indian Ocean (Recent). *Smithsonian Contributions to Paleobiology 43*, 1–66.

Cooper, G.A. 1981b: Brachipoda from the Gulf of Gascogne, France (Recent). *Smithsonian Contributions to Paleobiology 44*, 1–29.

Costa, O.G. 1851–1852: Brachiopods. *In: Fauna del Regno di Napoli ossia enumerazione di Autti gli Animale – contenente la descrizione de nuovio poco esattamente conosciuti – di Costa (continuata da A. Costa). Part 5*, 1–60.

Curry, G.B. 1982: Ecology and population structure of the Recent brachiopod *Terebratulina* from Scotland. *Palaeontology 25:2*, 227–246.

Curry, G.B. 1983: Ecology of the Recent rhynchonellid brachiopod *Cryptopora* from the Rockall Trough. *Paleogeography, Paleoclimatology and Paleoecology 44*, 93–102.

Dall, W.H. 1870: A revision of the Terebratulidae and Lingulidae. *American Journal of Conchology 6*, 88–168.

Dall, W.H. 1900: Some names which must be discarded. *Nautilus 14:4*, 44–45.

Davidson, T. 1852: A monograph of British Cretaceous Brachiopoda. Part 1. *Palaeontographical Society*, 5–54.

Davidson, T. 1854: A monograph of British Cretaceous Brachiopoda. Part 2. *Palaeontographical Society 20*, 55–177.

Davidson, T. 1874: A monograph of the British Fossil Brachiopoda. *Palaeontographical Society 4:1 (supplement)*, 1–72.

Deroo, G. 1966: Cytheracea (Ostracodes) du Maastrichtien de Maastricht (Pays-Bas) et des régions voisines; résultats stratigraphiques et paléontologiques de leur études. *Mededelingen van de Geologisches Stichting, Serie C 2*, 1–197.

Ekdale, A.A. & Bromley, R.G. 1984: Sedimentology and ichnology of the Cretaceous–Tertiary boundary in Denmark: Implications for the causes of the terminal Cretaceous extinction. *Journal of Sedimentary Petrology 54:3*, 681–703.

Felder, W.M. 1975: Lithostratigraphie van het Boven-Krijt en het Dano-Montien in Zuid-Limburg en het aangrenzende gebied. *Toelichting bij geologische overzichtskaarten van Nederland*, 63–72.

Fischer, P. & Oehlert, D.P. 1891: Brachiopods. *In: Expedition Scientifique du Travailleur et du Talisman (1880–1883)*. 139 pp. Paris.

Floris, S., Hansen, H.J., Håkansson, Krüger, J. & Surlyk, F. 1971: Geologi på øerne. *Varv ekskursionsfører 2*. 96 pp. Copenhagen.

Forchhammer, G. 1825: Om de Geognostiske Forhold i en Deel af Sjaelland og Naboeöerne. *Det Kongelige Danske Videnskabernes Selskab 2*.

Gealy, E.L., Winterer, E.L. & Moberly, R. 1971: Methods, conventions and general observations. *Initial Report of the Deep Sea Drilling Project 7*. 9–26.

Gould, S.J. 1977: *Ontogeny and Phylogeny*. 501 pp. Harvard University Press, Cambridge.

Gray, J.E. 1840: *Synopsis of the contents of the British Museum*, 42nd edit. 370 p. London.

Gray, J.E. 1848: On the arrangement of the Brachiopoda. *Annals & Magazine of Natural History 2:2*, 435–440.

Hagenow, F.v. 1842: Monographie der Rügen'schen Kreide-Versteinungen, Part 3: Mollusken. *Neues Jahrbuch für Mineralogie*, 528–575.

Håkansson, E., Perch-Nielsen, K. & Bromley, R.G. 1974: Maastrichtian chalk of NW-Europe – a pelagic shelf sediment. *Special Publications of the international Association of Sedimentology 1*, 211–233.

Håkansson, E. & Hansen, J,M. 1979: Guide to Maastrichtian and Danian boundary strata in Jylland. *In* Birkelund, T. & Bromley, R. (eds.), *Symposium on Cretaceous–Tertiary boundary events 1*, 171–188. Copenhagen.

Håkansson, E. & Thomsen, E. 1979: Distribution and types of bryozoan communities at the boundary in Denmark. *In* Birkelund, T. & Bromley, R. (eds.), *Symposium on Cretaceous–Tertiary boundary events 1*, 78–91. Copenhagen.

Hansen, J.M. 1977: Dinoflagellate stratigraphy and echinoid distribution in Upper Maastrichtian and Danian deposits from Denmark. *Bulletin of the Geological Society of Denmark 26*, 1–26.

Hansen, J.M. 1979a: Age of the Mo-clay Formation. *Bulletin of the Geological Society of Denmark 27*, 89–91.

Hansen, J.M. 1979b: A new dinoflagellate zone in the Maastrichtian/Danian boundary in Denmark. *The Geological Survey of Denmark, Yearbook, 1978*, 10–19.

Hansen, J.M. 1979c: Dinoflagellate zonation around the boundary. *In* Birkelund, T. & Bromley, R. (eds.), *Symposium on Cretaceous–Tertiary boundary events 1*, 136–141. Copenhagen.

Hatai, K. 1965: *In* Moore, R.C. (ed.) *Treatise on Invertebrate Paleontology (H) Brachiopoda 2*, H831–H832. Geological Society of America & University of Kansas Press, Lawrence.

Jablonski, D. & Bottjer, D.J. 1983: Soft bottom epifaunal suspension-feeding assemblages of the Late Cretaceous. *In* Tevesz, M.J. & McCall, P.L. (eds.): *Biotic Interactions in Recent and Fossil Benthic Communities*, 747–812. Plenum Press, New York.

Jablonski, D. & Lutz, R.A. 1980: Molluscan larval shell morphology: Ecological and paleontological applications. *In* Rhoads, D.C. & Lutz, R.A. (eds.): *Skeletal Growth of Aquatic Organisms*, 323–377. Plenum Press, New York.

Jackson, J.B.C., Goreau, T.F. & Hartman, W.D. 1971: Recent brachiopod–coralline sponge communities and their paleoecological significance. *Science 173*, 323–377.

Jeffreys, G. 1869: The deep-sea dredging expedition in H.M.S. *Porcupine*. 1: Natural History. *Nature 1:2*, 136.

Jeffreys, G. 1876: On some new and remarkable North-Atlantic Brachiopoda. *Annals and Magazine of Natural History 4:18*, 250–253.

[Johansen, M.B. 1982: *The Maastrichtian–Danian boundary in Denmark illustrated by the brachiopods at Nye Kløv, Northern Jylland, Denmark (systematics, paleoecology and evolution). Unpublished Thesis, University of Copenhagen.* 270 pp.]

[Johansen, M.B. 1986: *Colonisation, mass extinction and adaptive radiation – an evolutionary and stratigraphical study of the Upper Cretaceous–Lower Tertiary brachiopod fauna of northwestern Europe. Unpublished Ph.D. dissertation, University of Copenhagen.* 500 pp.]

Johansen, M.B. & Surlyk, F. 1983: The faunal turn-over of the brachiopods of Cretaceous–Tertiary boundary strata, Nye Kløv, Denmark. *15. Pacific Science Congress, University of Otago, Dunedin, New Zealand. Abstract*, 10–11.

King, W. 1859: On *Gwynia, Dielasma* and *Macandrevia*, three new genera of Palliobranchiata Mollusca. *Proceedings of the Dublin University Zoological Botanical Association 1:3*, 256–262.

Koenen, A.v. 1885: Ueber eine Paleocäne Fauna von Kopenhagen. *Abhandlungen der Königlichen Gesellschaft der Wissenschaft zu Göttingen*, 32 pp.

Kuhn, O. 1949: *Lehrbuch der Paläozoologie*. 326 pp. Schweizerbart, Stuttgart.

LaBarbera, M. 1981: The ecology of Mesozoic *Gryphaea, Exogyra* and *Ilymatogyra* (Bivalvia: Mollusca) in a modern ocean. *Paleobiology 7*, 510–526.

Logan, A. 1975: Ecological observations on the Recent articulate brachiopod *Argyrotheca bermudana* Dall from the Bermuda Platform. *Bulletin of Marine Sciences 25*, 186–204.

Logan, A. 1979: The recent Brachiopoda of the Mediterranean Sea. *Bulletin of the Institute of Oceanography, Monaco 72:1434*, 1–112.

Lutherbacher, H.P. & Premoli Silva, I. 1966: Biostratigraphica del limite Cretaceo–Terziario nell'Appennino centrale. *Rivista Italiana Paleontologia e Stratigrafia 70*, 67–128.

Martini, E. 1971: Standard Tertiary and Quaternary calcareous nannoplankton zonation. *Proceedings of the 2nd Planktonic Conference*, 739–785. Rome.

McCammon, H.M. 1970: Variation in Recent brachiopod populations. *Bulletin of the Geological Institutions of the University of Upsala 2*, 41–48.

Moore, R.C. (ed.) 1965: *Treatise on Invertebrate Paleontology (H) Brachiopoda. 927 pp. Geological Society of America & University of Kansas Press*, Lawrence.

de Morgan, J. 1883: Note sur quelques espèces nouvelles de Megathyrides. *Bulletin de la Societé Zoologique de France 7*, 371–396.

Muir-Wood, H. 1955: *A history of the classification of the phylum Brachiopoda*. 124 pp. British Museum (Natural History), London.

Muir-Wood, H. 1959: Report on the Brachiopoda of the John Murray Expedition. *John Murray Expedition 1933–34, Scientific Report 10:6*, 283–317. British Museum (Natural History), London.

Muir-Wood, H. 1965: Mesozoic and Cenozoic Terebratulidina. *In* Moore, R.C. (ed.): *Treatise on Invertebrate Paleontology (H) Brachiopoda 2*, H762–H864. Geological Society of America & University of Kansas Press, Lawrence.

Naidin, D.P. 1979: The Cretaceous–Tertiary boundary in the USSR. *In* Christensen, W.K. & Birkelund, T. 1979 (eds.): *Symposium on Cretaceous–Tertiary boundary events 2*. 188–202. Copenhagen.

Nielsen, K.B. 1909: Brachiopoderni i Danmarks Kridtaflejringer. *Det Kongelige Danske Videnskabernes Selskab 7:4*, 129–178.

Nielsen, K.B. 1911: Brachiopoderne i Faxe. *Meddelelser Dansk Geologisk Forening 3*, 599–618.

Nielsen, K.B. 1914: Some remarks on the Brachiopods in the Chalk of Denmark. *Meddelelser Dansk Geologisk Forening 4*, 287–296.

Nielsen, K.B. 1920: Inddelingen af Danien'et i Danmark og Skåne. *Meddelelser Dansk Geologisk Forening 5:19*, 1–16.

Nielsen, K.B. 1921: Nogle Bemærkninger om de store Terebratler i Danmarks Kridt og Danien Aflejringer. *Meddelelser Dansk Geologisk Forening 6:3*, 2–18.

Nielsen, K.B. 1928: *Argiope*-arterne i danske aflejringer. *Meddelelser Dansk Geologisk Forening 7*, 215–226.

Nielsen, K.B. 1937: Faunaen i ældre Danium ved Korporalskroen. *Meddelelser Dansk Geologisk Forening 9*, 117–162.

Noble, J.P.A. & Logan, A. 1981: Size-frequency distributions and taphonomy of brachiopods: a recent model. *Paleogeography, Paleoclimatology and Paleoecology 36*, 87–105.

Ødum, H. 1926: Studier over Danien i Jylland og på Fyn. *Danmarks Geologiske Undersøgelser 2:45*, 1–306.

d'Orbigny, A. 1847: Considérations zoologiques et géologiques sur les brachiopodes ou palliobranches. *Académie Scientifique Paris Comptes Rendus 25*, 193–195, 266–269.

Owen, E.F. 1970: A revision of the brachiopod subfamily Kingeninae Elliott. *Bulletin of the British Museum (Natural History) 19:2*, 29–83.

Owen, E.F. 1977: Evolutionary trends in some Mesozoic Terebratellacea. *Bulletin of the British Museum (Natural History) Geology 28:3*, 205–253.

Paine, R.T. 1970: The sediment occupied by Recent lingulid brachiopods and some paleoecological implications. *Paleogeography, Paleoclimatology and Paleoecology 7*, 21–31.

Panow, E. 1969: Contribution to the knowledge of the Brachiopods from the Upper Cretaceous of the Krakow district. *Rocznik Polskiego Towarzystwa Geologicznego Krakow 39:4*, 555–608. (Posthumous publication, prepared by Biernat, G. and Popiel-Barczyk, E.)

Peake, N.B. & Hancock, J.M. 1961: The Upper Cretaceous of Norfolk. *Transactions of the Norfolk & Norwich Naturalists' Society 19*, 293–339.

Perch-Nielsen, K. 1979a: Calcareous nannofossils in Cretaceous–Tertiary boundary sections in Denmark. *In* Christensen, W.K. & Birkelund, T. 1979 (eds.): *Symposium on Cretaceous–Tertiary boundary events 2*, 120–126. Copenhagen.

Perch-Nielsen, K. 1979b: Calcareous nannofossil zonation at the Cretaceous–Tertiary boundary in Denmark. *In* Birkelund, T. & Bromley, R. (eds.), *Symposium on Cretaceous–Tertiary boundary events 1*, 115–135. Copenhagen.

Pettitt, N.E. 1950: A monograph of the Rhynchonellidae of the British chalk. *Palaeontographical Society pt. 1–2*, 1–52.

Popiel-Barczyk, E. 1968: Upper Cretaceous Terebratulids (Brachiopoda) from the Middle Vistula Gorge. *Prace Museum Ziemi 12*, 1–83.

Popiel-Barczyk, E. 1973: Albian–Cenomanian brachiopods from the environs of Annopol on the Vistula with some remarks on related species from Cracow Region. *Prace Museum Ziemi 2*, 119–150.

Popiel-Barczyk, E. 1977: A further study of Albian–Cenomanian brachiopods from the environs of Annopol on the Vistula with some remarks on related species from the Cracow Region, Poland. *Prace Museum Ziemi 26*, 25–54.

Posselt, H.J. 1894: Brachiopoderne i den danske Kridtformation. *Danmarks geologiske Undersøgelser 2:4*, 1–59.

Rasmussen, H.W. 1961: A monograph on the Cretaceous Crinoidea. *Det Kongelige Danske Videnskabernes Selskab, Biologiske Skrifter 12*, 1–428.

Rasmussen, H.W. 1965: The Danian affinities of the tuffeau de Ciply in Belgium and the 'post-Maastrichtian' in the Netherlands. *Mededelingen van de Geologisches Stichting, Serie C 17*, 33–50.

Ravn, J.P.J. 1903: Molluskerne i Danmarks Kridtaflejringer. *Det Kongelige Danske Videnskabernes Selskab, Biologiske Skrifter 11:6*, 335–446.

Rhoads, D.C. 1970: Mass properties, stability and ecology of marine muds related to burrowing activity. *In* Crimes, T.P. & Harper, J.C. (eds.): *Trace Fossils*, 391–406. Seel House Press, Liverpool.

Rhoads, D.C. 1974: Organism–sediment relations on the muddy seafloor. *Annual Reviews of Oceanography and Marine Biology 12*, 263–300.

Richards, R.P. & Bambach, R.K. 1975: Population dynamics of some Paleozoic brachiopods and their paleoecological significance. *Journal of Paleontology 49*, 775–798.

Richardson, J. 1981a: Brachiopods and pedicles. *Paleobiology 7:1*, 87–95.

Richardson, J. 1981b: Distribution, orientation and movement in six species of New Zealand articulate brachiopods. *New Zealand Journal of Zoology 8:2*, 189–196.

Richardson, J. 1981c: Brachiopods in mud: Resolution of a dilemma. *Science 211*, 1161–1163.

Richardson, J. 1986: Brachiopods. *Scientific American 255:3*, 96–102.

Roemer, F.A. 1841: *Die Versteinerungen des norddeutschen Kreidegebirges.* 145 pp. Hannover.

Romein, A.J.T. 1977: Calcareous nannofossils from the Cretaceous–Tertiary boundary interval in the Barranco des Gredero (Caravaca, SE. Spain). *Proceedings Koninklijke Nederlandse Akademie van Wetenschappen 80:3*, 256–279.

Rosenkrantz, A. 1924: Nye Iagttagelser over Cerithiumkalken i Stevns Klint med Bemærkninger om Grænsen mellem Kridt og Tertiær. *Meddelelser Dansk Geologisk Forening 6*, 28–31.

Rosenkrantz, A. 1937: Bemærkninger om det Østsjællandske Daniens Stratigrafi og Tektonik. *Meddelelser Dansk Geologisk Forening 9*, 199–212.

Rosenkrantz, A. 1940: Faunaen i Cerithiumkalken og det hærdnede Skrivekridt i Stevns Klint. *Meddelelser Dansk Geologisk Forening 12*, 509–514.

Rudwick, M.J.S. 1961: The anchorage of articulate brachiopods on soft substrata. *Palaeontology 4*, 475–476.

Rudwick, M.J.S. 1962: Notes on the ecology of brachiopods in New Zealand. *Transactions of the Royal Society of New Zealand Zoology 1:25*, 327–335.

Rudwick, M.J.S. 1965: Ecology and Paleoecology. *In* Moore, R.C. (ed.): *Treatise on Invertebrate Paleontology (H) Brachiopoda 1*, H199–H213. Geological Society of America & University of Kansas Press, Lawrence.

Rudwick, M.J.S. 1970: *Living and Fossil Brachiopods.* 199 pp. Hutchinson University Library, London.

Sahni, M.R. 1925: Morphology and zonal distribution of some Chalk Terebratulids. *Annals & Magazine of Natural History 9:15*, 353–385.

Scacchi, A. & Philippi, R.A. 1844: *Enumeratio Molluscorum Siciliae 2*, pp. 1–303.

Schloenbach, U. 1866: Beiträge zur Paläontologie der Jura- und Kreide Formation im NW-lichen Deutschland. St. II Kritische Studien über Kreide-Brachiopoden. *Paläontographica 12*, 1–66.

Schlottheim, E.F.v. 1813: Beiträge zur Naturgeschichte der Versteinerungen in geognostischer Hinsicht. *Taschenbuch Gesteinische Minerale.* Erste Abteilung, 3–134. Frankfurt am Main.

Scholle, P.A. 1977: Current oil and gas production from North American Upper Cretaceous chalks. *Circular of the Geological Survey*

of the United States 767, 51 pp.

Skelton, P.W. 1979: Gregariousness and proto-cooperation in rudists (Bivalvia). *In* Larwood, G.P. & Rosen, B.R. (eds.): *Biology and Systematics of Colonial Organisms*, 257–279. Academic Press, New York.

Smit, J. 1982: Extinction and evolution of planktonic foraminifera after a major impact at the Cretaceous–Tertiary boundary. *Geological Society of America Special Papers 190*, 329–352.

Smit, J. & Hertogen, J. 1980: An extraterrestrial event at the Cretaceous–Tertiary boundary. *Nature 285*, 198–200.

Smit, J. & Romein, A.J.T. 1985: A sequence of events across the Cretaceous–Tertiary boundary. *Earth & Planetary Science Letters 74*, 155–170.

Sohl, N.F. & Koch, C.F. 1982: Substrate preference among some Late Cretaceous shallow-water benthic molluscs. *Geological Society of America, Abstract with Programs 14*, 84.

Sowerby, J. 1816: *The mineral conchology of Great Britain 2*, 235 pp. London.

Sowerby, J. 1821: *The mineral conchology of Great Britain 3*, 218 pp. London.

Stanley, S.M. 1970: Relation of shell form to life habits in the Bivalvia (Mollusca). *Geological Society of America Memoirs 125*, 296 pp.

Stanley, S.M. 1979: *Macroevolution: Pattern and Process.* 332 pp. Freeman, San Francisco.

Steinich, G. 1963a: Fossile Spicula bei Brachiopoden der Rügener Schreibkreide. *Geologie 12:5*, 604–610.

Steinich, G. 1963b: Drei neue Brachiopodengattungen der Subfamilie Cancellothyridinae Thomson. *Geologie 12:6*, 732–740.

Steinich, G. 1963c: Zur Morphogenese des Foramens der Rhynchonellida. *Geologie 12:10*, 1204–1209.

Steinich, G. 1965: Die artikulaten Brachiopoden der Rügener Schreibkreide (Unter-Maastricht.) *Paläontologische Abhandlungen 2:1*, 1–200.

Steinich, G. 1967: Neue Brachiopoden aus der Rügener Schreibkreide (Unter-Maastricht.). I. Draciinae – eine neue Unterfamilie der Cancellothyrididae Thomson. *Geologie 16:10*, 1145–1155.

Steinich, G. 1968a: Neue Brachiopoden aus der Rügener Schreibkreide (Unter-Maastricht.). II. Die Platidiidae Thomson. *Geologie 17:2*, 192–209.

Steinich, G. 1968b: Neue Brachiopoden aus der Rügener Schreibkreide (Unter-Maastricht.). III. *Dalligas nobilis* gen. et sp.nov. und *Kingena* sp. *Geologie 17:3*, 336–347.

Stenzel, H.B. 1971: Oysters. *In* Moore, R.C. (ed.): *Treatise on Invertebrate Paleontology (N) Mollusca 6:3*, N953–N1224. Geological Society of America & University of Kansas Press, Lawrence.

Stewart, I.R. 1981: Population structure of articulate brachiopod species from soft and hard substrates. *New Zealand Journal of Zoology 8*, 197–207.

[Surlyk, F. 1969: *A study on the articulate brachiopods of the Danish White Chalk. (U. Campanian and Maastrichtian) with a review of the sedimentology of the White Chalk and the flora and fauna of the Chalk Sea.* Unpublished prize dissertation. University of Copenhagen. 319 pp.]

Surlyk, F. 1970a: Two new brachiopods from the Danish White Chalk (Maastrichtian). *Bulletin of the Geological Society of Denmark 20*, 152–161.

Surlyk, F. 1970b: Die Stratigraphie des Maastricht von Dänemark und Norddeutschland aufgrund von Brachiopoden. *Newsletters of Stratigraphy 12*, 7–16.

Surlyk, F. 1971: Skrivekridtklinterne på Møn. *In* Floris, S., Hansen, H.J., Håkansson, E., Krüger, J. & Surlyk, F. (eds.): Geologi på Øerne. *Varv ekskursionsfører 2.* 4–24, Copenhagen.

Surlyk, F. 1972: Morphological adaptions and population structures of the Danish Chalk brachiopods (Maa., U.Cret.). *Det Kongelige Danske Videnskabernes Selskab, Biologiske Skrifter 19:2*, 57 pp.

Surlyk, F. 1973: Autecology and taxonomy of two Upper Cretaceous craniacean brachiopods. *Bulletin of the Geological Society of Denmark 22*, 219–243.

Surlyk, F. 1974: Life habit, feeding mechanism and population structure of the Cretaceous brachiopod genus *Aemula*. *Paleogeography, Paleoclimatology and Paleoecology 15*, 185–203.

Surlyk, F. 1979a: Maastrichtian brachiopods from Denmark. *In* Birkelund, T. & Bromley, R. (eds.): *Symposium on Cretaceous–Tertiary boundary events 1*, 45–50. Copenhagen.

Surlyk, F. 1979b: Guide to Stevns Klint. *In* Birkelund, T. & Bromley, R. (eds.), *Symposium on Cretaceous–Tertiary boundary events 1*, 164–170. Copenhagen.

Surlyk, F. 1982: Brachiopods from the Campanian–Maastrichtian boundary sequence, Kronsmoor (NW Germany). *Geologisches Jahrbuch A61*, 259–277.

Surlyk, F. 1983: The Maastrichtian stage in NW Europe, and its brachiopod zonation. *In* Birkelund, T., Bromley, R., Christensen, W.K., Håkansson, E. & Surlyk, F. (eds.): *Cretaceous Stage Boundaries*, 191–196. Copenhagen.

Surlyk, F. 1984: The Maastrichtian Stage in NW Europe, and its brachiopod zonation. *Bulletin of the Geological Society of Denmark 33:1–2*, 217–224.

Surlyk, F. & Birkelund, T. 1977: An integrated stratigraphical study of fossil assemblages from the Maastrichtian White Chalk of NW Europe. *In* Kaufmann, E.G. & Hazel, J.E. (eds.): *Concepts and Methods of Biostratigraphy*, 259–281. Dowden, Hutchinson & Ross, Stroudsburg, Pennsylvania.

Surlyk, F. & Johansen, M.B. 1981: Extinction pattern of late Cretaceous brachiopods compatible with catastrophic change of the marine calcareous shelled biota. *Abstract to symposium on the Cretaceous–Tertiary Boundary, American Association of Advanced Sciences*. January 1982.

Surlyk, F. & Johansen, M.B. 1984: End-Cretaceous brachiopod extinctions in the chalk of Denmark. *Science 223*, 1174–1177.

Tappan, H. & Loeblich, A.R.Jr. 1971: Geologic implications of fossil phytoplankton evolution and time-space distribution. *Geological Society of America Special Papers 127*, 247–339.

Taylor, J.D. 1981: The evolution of predators in the Late Cretaceous and their ecological significance. *In* Forey, P.L. (ed.): *The Evolving Biosphere*, 229–240. British Museum (Natural History) and Cambridge University Press, London.

Thayer, C.W. 1975a: Morphological adaptations of benthic invertebrates to soft substrata. *Journal of Marine Research 33*, 177–189.

Thayer, C.W. 1975b: Strength of pedicle attachment in articulate brachiopods: ecological and palecologic significance. *Paleobiology 1*, 388–399.

Thayer, C.W. 1979: Biological bulldozers and the evolution of marine benthic communities. *Science 203*, 458–461.

Thayer, C.W. & Steele-Petrovic, H.M. 1975: Burrowing of the lingulid brachiopod *Glottidia pyramidata*: its ecologic and paleoecologic significance. *Lethaia 8*, 209–221.

Thierstein, H.R. 1981: Late Cretaceous nannoplankton and the change at the Cretaceous–Tertiary boundary. *Society of Economic Paleontologists and Mineralogists Special Publication 32*, 335–394.

Thierstein, H.R. 1982: Terminal Cretaceous plankton extinctions: A critical assessment. *Geological Society of America Special Papers 190*, 385–399.

Thomson, A.J. 1926: A revision of the subfamilies of the Terebratulidae (Brachiopoda). *Annals & Magazine of Natural History 9:18*, 523–530.

Thomson, A.J. 1927: Brachiopod morphology and genera (Recent and Tertiary). *New Zealand Board of Science & Art Manual 7*, 1–329.

Titova, M.V. 1977: Pozdnemelovye Cancellothyrididae (Brachiopoda) Turkmenii. [Late Cretaceous Cancellothyrididae (Brachiopoda) of Turkmenistan.] *Paleontologicheskij Zhurnal 4*, 73–85.

Troelsen, J. 1937: Om den stratigrafiske inddeling af skrivekridtet i Danmark. *Meddelelser Dansk Geologisk Forening 9*, 260–263.

Valentine, J.W. 1974: *Evolutionary paleoecology of the marine biosphere*. 511 pp. Prentice Hall. New York.

Vermeij, G.J. 1977: The Mesozoic marine revolution: Evidence from snails, predators and grazers. *Paleobiology 3*, 245–258.

Vermeij, G.J. 1978: *Biogeography and Adaptation: Patterns of Marine Life*. 332 pp. Harvard University Press, Cambridge, Mass.

Waagen, W.H. 1883: Salt Range Fossils, Part 4(2) Brachiopoda: *Memoirs of the Geological Survey of India. Palaeontologia Indica 13:1*, 391–546.

Williams, A. & Rowell, A.J. 1965: Brachiopod anatomy and morphology. *In* Moore, R.C. (ed.): *Treatise on Invertebrate Paleontology (H) Brachiopoda 1*, H6–H155. Geological Society of America & University of Kansas Press, Lawrence.

Wind, J. 1953: Kridtaflejringer i Jylland. *Natur og Museum 59*, 73–84.

Woodward, S.P. 1833: *Outline of the Geology of Norfolk*. 1–54. Norwich.

Plates

Plate 1

Cretirhynchia sp., *Carneithyris* sp., and *Neoliothyrina?* sp.

☐ 1. *Cretirhynchia* sp. Interior of fragmented juvenile ventral valve. Sample NK 23, Lower Danian. MGUH 16896.

☐ 2. *Carneithyris* sp. Interior of juvenile ventral valve. Sample NK 26, Lower Danian. MGUH 16897A.

☐ 3. *Carneithyris* sp. Interior of juvenile dorsal valve from the same specimen as in 2 showing crura. Sample NK 26, Lower Danian. MGUH 16897B.

☐ 4. *Carneithyris* sp. Complete small juvenile in dorsal view. Sample NK 25, Lower Danian. MGUH 16898.

☐ 5. *Neoliothyrina?* sp. Interior of juvenile fragmented dorsal valve showing cardinalia. Sample NK 24, Lower Danian. MGUH 16899.

☐ 6. *Neoliothyrina?* sp. Interior of adult, fragmented dorsal valve showing part of cardinalia. Sample NK 24, Lower Danian. MGUH 16900.

Scale bars 0.5 mm.

Plate 2

Cretirhynchia sp., *Carneithyris subcardinalis* Sahni 1925, and *Meonia semiglobularis* (Posselt 1894).

☐ 1. *Cretirhynchia* sp. Posterior part of adult specimen showing characteristic collar-like deltidial plates. Sample NK1, Upper Maastrichtian. MGUH 16901.

☐ 2. *Carneithyris subcardinalis*. Complete juvenile specimen in dorsal view. Sample NK 4, Upper Maastrichtian. MGUH 16902.

☐ 3. *Meonia semiglobularis*. Complete adult specimen. Sample NK 8, Maastrichtian–Danian boundary clay. MGUH 16903. ☐ A. Dorsal view. ☐ B. Oblique lateral view showing hemisphaerical shell. ☐ C. Enlargement of beak.

☐ 4. *Meonia semiglobularis*. Complete medium-sized specimen. Sample NK 8, Maastrichtian–Danian boundary clay. MGUH 16904. ☐ A. Dorsal view. ☐ B. Oblique lateral view.

☐ 5. *Meonia semiglobularis*. Complete juvenile specimen. Sample NK 8, Maastrichtian–Danian boundary clay. MGUH 16905. ☐ A. Dorsal view. ☐ B. Oblique lateral view showing flattened biconvex shell.

Scale bars 0.5 mm except where otherwise stated.

Plate 3

Cryptopora perula n.sp. Figs. 1–3 show ontogenetic development.

☐ 1. Holotype. Complete adult specimen. Sample NK 24, Lower Danian. MGUH 16906. ☐ A. Dorsal view. ☐ B. Lateral view.

☐ 2. Complete juvenile in dorsal view. Sample Nk 23, Lower Danian. MGUH 16907.

☐ 3. Complete small juvenile specimen in dorsal view. Sample NK 23, Lower Danian. MGUH 16908.

☐ 4. Posterior part of fragmented ventral valve showing foramen and hinge teeth. Sample NK 25, Lower Danian. MGUH 16909.

☐ 5. Complete adult specimen with posterior part of ventral removed to show crura, crural processes and descending branches. Locality 'Dania', sample Dania D 106 B(1) (MBJ) from Lower the Danian, around 3,0 m above the Maastrichtian–Danian boundary. MGUH 16910.

☐ 6. Interior of juveline dorsal valve showing crura and recrystallized schizolophe(?). Sample NK 27, Lower Danian. MGUH 16911.

☐ 7. Interior of framented dorsal valve showing median septum. Sample NK 25, Lower Danian. MGUH 16912.

Scale bars 0.5 mm except where otherwise stated.

Plate 4

Terebratulina chrysalis (Schlottheim 1813), and *Terebratulina* cf. *longicollis* Steinich 1965.

☐ 1. *Terebratulina chrysalis*. Interior of medium-sized dorsal valve showing ring-shaped brachidium and part of recrystallized plectolophe. Sample NK 30, Lower Danian. MGUH 16913.

☐ 2. *Terebratulina chrysalis*. Interior of large, adult dorsal valve showing brachidium. Sample NK 30, Lower Danian. MGUH 16914.

☐ 3. *Terebratulina chrysalis*. Interior of juvenile dorsal valve showing brachidium and part of recrystallized plectolophe. Sample NK 30, Lower Danian. MGUH 16915.

☐ 4. *Terebratulina chrysalis*. Complete small juvenile specimen in dorsal view. Sample NK 30, Lower Danian. MGUH 16916.

☐ 5. *Terebratulina chrysalis*. Complete juvenile specimen in dorsal view. Sample NK 30, Lower Danian. MGUH 16917.

☐ 6. *Terebratulina* cf. *longicollis*. Sample NK 28, Lower Danian. MGUH 16918A. ☐ A. Interior of adult dorsal valve showing brachidium. ☐ B. Interior of adult ventral valve of the same specimen.

☐ 7. *Terebratulina* cf. *longicollis*. Complete juvenile specium in dorsal view. Sample NK 27, Lower Danian. MGUH 16919.

Scale bars 0.5 mm.

Plate 5

Terebratulina aff. *rigida* (Sowerby 1821).

☐ 1. Complete adult specimen. Sample NK 26, Lower Danian. MGUH 16920. ☐ A. Dorsal view. ☐ B. Lateral view. ☐ C. Specimen in A opened. Interior of dorsal valve showing complete brachidium and almost complete recrystallized plectolophe. MGUH 16920A. ☐ D. Interior of ventral valve showing ventral part of recrystallized plectolophe. MGUH 16920B. ☐ E. Specimen in C in oblique lateral view.

☐ 2. Complete juvenile specimen in dorsal view. Sample NK 26, Lower Danian. MGUH 16921.

☐ 3. Interior of juvenile dorsal valve showing fragmented brachidium and recrystallized spicular skeleton. Sample NK 26, Lower Danian. MGUH 16922.

☐ 4. Complete small juvenile specimen in dorsal view. Sample NK 26, Lower Danian. MGUH 16923.

Scale bars 0.5 mm.

Plate 6

Terebratulina kloevensis n.sp. Figs 1–4 show ontogenetic development.

□ 1. Holotype. Complete adult specimen. Sample NK 30, Lower Danian. MGUH 16924. □ A. Dorsal view. □ B. Lateral view.

□ 2. Adult specimen in ventral view. Shell is opened to show ring-formed brachidium. Sample NK 29, Lower Danian. MGUH 16925.

□ 3. Complete medium-sized specimen in dorsal view. Sample NK 30, Lower Danian. MGUH 16926.

□ 4. Complete juvenile specimen in dorsal view sample NK 30, Lower Danian. MGUH 16927.

□ 5. Exterior of adult vental valve showing rib pattern and growth. By comparison with, e.g., the specimen in 1A, the difference in sculpture on dorsal and ventral valves is illustrated. Sample NK 29. Lower Danian. MGUH 16928.

Scale bars 0.5 mm.

Plate 7

Rugia acutirostis Steinich 1965, *Rugia tenuicostata* Steinich 1963, *Terebratulina faujasii* (Roemer 1841), *Terebratulina longicollis* Steinich 1965, and *Terebratulina gracilis* (Schlottheim 1813). Figs. 7–9 show ontogenetic development of *Terebratulina gracilis*.

□ 1. *Rugia acutirostis*. Complete medium-sized specimen. Sample NK 8. Maastrichtian–Danian boundary clay. MGUH 16929. □ A. Dorsal view. □ B. Enlargement of posterior part showing a very acute beak and a small foramen.

□ 2. *Rugia tenuicostata*. Slightly damaged adult dorsal valve in exterior view. Sample NK 4, Upper Maastrichtian. MGUH 16930.

□ 3. *Terebratulina faujasii*. Complete adult specimen in dorsal view. Sample NK 8. Maastrichtian–Danian boundary clay. MGUH 16931.

□ 4. *Terebratulina longicollis*. Complete adult specimen in dorsal view. Sample NK 5, Upper Maastrichtian. MGUH 16932.

□ 5. *Terebratulina longicollis*. Complete medium-sized specimen in dorsal view. Sample NK 5, Upper Maastrichtian. MGUH 16933.

□ 6. *Terebratulina longicollis*. Complete juvenile specimen in dorsal view. Sample NK 5, Upper Maastrichtian. MGUH 16934.

□ 7. *Terebratulina gracilis*. Complete small juvenile specium in dorsal view. Sample NK 6, Upper Maastrichtian. MGUH 16935.

□ 8. *Terebratulina gracilis*. Complete juvenile specimen. Sample NK 6, Upper Maastrichtian. MGUH 16936. □ A. Dorsal view. □ B. Oblique lateral view showing biconvex shell.

□ 9. *Terebratulina gracilis*. Fragmented adult specimen. Sample NK 5, Upper Maastrichtian. MGUH 16937. □ A. Dorsal view. □ B. Enlargement of posterior part showing incurved beak and permesothyridid forament. □ C. Oblique lateral view.

Scale bars 0.5 mm, except where otherwise stated.

Plate 8

Terebratulina kloevensis n.sp., *Terebratulina* aff. *rigida* (Sowerby 1821), and *Gisilina jasmundi* Steinich 1965.

□ 1. *Terebratulina kloevensis*. Enlargement of anterior part of vental valve figured on Plate 6:5 showing growth pattern and sculpture of ribs. Sample NK 29. Lower Danian. MGUH 16928.

□ 2. *Terebratulina* aff. *rigida*. Enlargement of anterio-dextrally part of vental valve figured on Plate 5:1A showing growth pattern and sculpture of ribs. Sample NK 26, Lower Danian. MGUH 16920.

□ 3. *Gisilina jasmundi*. Exterior of adult dorsal valve. Sample NK 5, Upper Maastrichtian. MGUH 16938.

□ 4. *Terebratulina kloevensis*. Interior of adult ventral valve. Sample NK 29, Lower Danian. MGUH 16939.

□ 5. *Terebratulina kloevensis*. Interior of adult dorsal valve showing cardinalia and base of crura. Sample NK 29, Lower Danian. MGUH 16940.

Scale bars 0.5 mm.

Plate 9

Rugia flabella n.sp. Figs. 1–4 show ontogenetic developemnt.

☐ 1. Holotype. Complete adult specimen. Sample NK 19, Lower Danian. MGUH 16941. ☐ A. Dorsal view. ☐ B. Lateral view.

☐ 2. Complete medium-sized specimen in dorsal view. Sample NK 24, Lower Danian. MGUH 16942.

☐ 3. Complete juvenile specimen in dorsal view. Sample NK 19, Lower Danian. MGUH 16943.

☐ 4. Complete small juvenile in dorsal view. Sample NK 19, Lower Danian. MGUH 16944.

☐ 5. Interior of adult ventral valve. Sample NK 24, Lower Danian. MGUH 16945.

☐ 6. Interior of adult dorsal valve showing cardinalia and almost complete brachidium. Sample NK 21, Lower Danian. MGUH 16946.

☐ 7. Interior of adult ventral valve showing recrystallized zygolophe(?). The recrystallized filaments of the lophophore are marked with ink. Sample NK 27, Lower Danian. MGUH 16947.

Scale bars 0.5 mm.

Plate 10

Rugia latronis n.sp. and *Rugia* sp. Figs. 1–3 show ontogenetic development of *Rugia latronis*.

□ 1. *Rugia latronis* n.sp. Holotype. Complete adult specimen. Sample NK 26, Lower Danian. MGUH 16948. □ A. Dorsal view. □ B. Enlargement of anterior sinistral part, showing rib growth pattern and sculpture.

□ 2. *Rugia latronis* n.sp. Complete medium-sized specimen in dorsal view. Sample NK 23, Lower Danian. MGUH 16949.

□ 3. *Rugia latronis* n.sp. Complete juvenile specimen in dorsal view. Sample NK 25, Lower Danian. MGUH 16950.

□ 4. *Rugia* sp. Complete medium-sized specimen in dorsal view. Sample NK 24, Lower Danian. MGUH 16951.

□ 5. *Rugia* sp. Interior of medium-sized ventral valve. Sample NK 24, Lower Danian. MGUH 16952.

Scale bars 0.5 mm except where otherwise stated.

Plate 11

Gwyniella persica n.sp. and *Argyrotheca* aff. *bronnii* (Roemer 1841). Figs. 1, 4, 5 show ontogenetic development of *Gwyniella persica*.

☐ 1. *Gwyniella persica*. Holotype. Slightly damaged two-valved adult specimen in dorsal view. Sample NK 26, Lower Danian. MGUH 16953.

☐ 2. *Gwyniella persica*. Interior of adult ventral valve. Sample NK 26, Lower Danian. MGUH 16954.

☐ 3. *Gwyniella persica*. Interior of adult dorsal valve showing median septum and cardinalia. Sample NK 19, Lower Danian. MGUH 16955.

☐ 4. *Gwyniella persica*. Complete medium-sized specimen in dorsal view. Sample NK 26, Lower Danian. MGUH 16956.

☐ 5. *Gwyniella persica*. Complete juvenile specimen in dorsal view. Sample NK 26, Lower Danian. MGUH 16957.

☐ 6. *Argyrotheca* aff. *bronnii*. Interior of juvenile dorsal valve. Sample NK 23, Lower Danian. MGUH 16958.

☐ 7. *Argyrotheca* aff. *bronnii*. Complete medium-sized specimen in dorsal view. Sample NK 23, Lower Danian. MGUH 16959.

Scale bars 0.5 mm except where otherwise stated.

Plate 12

Argyrotheca aff. *bronnii* (Roemer 1841), *Argyrotheca danica* (de Morgan 1883), *Argyrotheca coniuncta* Steinich 1965, and *Argyrotheca stevensis* (Nielsen 1928).

☐ 1. *Argyrotheca* aff. *bronnii*. Interior of adult dorsal valve. Sample NK 23, Lower Danian. MGUH 16960.

☐ 2. *Argyrotheca danica*. Complete juvenile specimen in dorsal view. Sample NK 5, Upper Maastrichtian. MGUH 16961.

☐ 3. *Argyrotheca danica*. Interior of fragmented adult dorsal valve. Sample NK 4, Upper Maastrichtian. MGUH 16962.

☐ 4. *Argyrotheca danica*. Interior of fragmented adult dorsal valve. Sample NK 4, Upper Maastrichtian. MGUH 16963.

☐ 5. *Argyrotheca coniuncta*. Adult dorsal valve. Sample NK 8, Maastrichtian–Danian boundary clay. MGUH 16964. ☐ A. Interior. ☐ B. Oblique lateral view showing median septeum in profile.

☐ 6. *Argyrotheca stevensis*. Adult dorsal valve. Sample NK 28, Lower Danian. MGUH 16965. ☐ A. Interior. ☐ B. Oblique lateral view showing profile of median septum.

Scale bars 0.5 mm except where otherwise stated.

Plate 13

Argyrotheca stevensis (Nielsen 1928) and *Argyrotheca* aff. *stevensis.*

☐ 1. *Argyrotheca stevensis* (Nielsen 1928). Complete adult specimen in dorsal view. Sample NK 26, Lower Danian. MGUH 16966.

☐ 2. *Argyrotheca stevensis* (Nielsen 1928). Interior of adult ventral valve. Sample NK 24, Lower Danian. MGUH 16967.

☐ 3. *Argyrotheca stevensis* (Nielsen 1928). Interior of juvenile ventral valve. Sample NK 26, Lower Danian. MGUH 16968.

☐ 4. *Argyrotheca stevensis* (Nielsen 1928). Interior of adult dorsal valve. Sample NK 26, Lower Danian. MGUH 16969.

☐ 5. *Argyrotheca stevensis* (Nielsen 1928). Complete medium-sized specimen in dorsal view. Dorsal valve contains boring produced by predaceous gastropod. Sample NK 23, Lower Danian. MGUH 16970.

☐ 6. *Argyrotheca* aff. *stevensis* (Nielsen 1928). Complete juvenile specimen in dorsal view. Sample NK 22, Lower Danian. MGUH 16971.

☐ 7. *Argyrotheca* aff. *stevensis* (Nielsen 1928). Complete adult specimen in dorsal view. Sample NK 22, Lower Danian. MGUH 16972.

Scale bars 0.5 mm except where otherwise stated.

Plate 14

Argyrotheca hirundo (Hagenow 1842). Figs. 1, 3, 4, 5, 8 show ontogenetic development and variation.

☐ 1. Complete adult specimen in dorsal view. Sample NK 26, Lower Danian. MGUH 16973.

☐ 2. Interior of adult dorsal valve. Sample NK 4, Upper Maastrichtian. MGUH 16974.

☐ 3. Complete medium-sized specimen in dorsal view. Sample NK 23, Lower Danian. MGUH 16975.

☐ 4. Complete juvenile specimen in dorsal view. Sample Nk 23, Lower Danian. MGUH 16976.

☐ 5. Complete small juvenile specimen in dorsal view. Sample NK 23, Lower Danian. MGUH 16977.

☐ 6. Interior of juvenile dorsal valve. Hair-like object on top of the picture does not belong to the specimen. Sample NK 23, Lower Danian. MGUH 16978.

☐ 7. Interior of medium-sized dorsal valve. Sample NK 23, Lower Danian. MGUH 16979.

☐ 8. Complete adult specimen in dorsal view. Sample NK 23, Lower Danian. MGUH 16980.

Scale bars 0.5 mm.

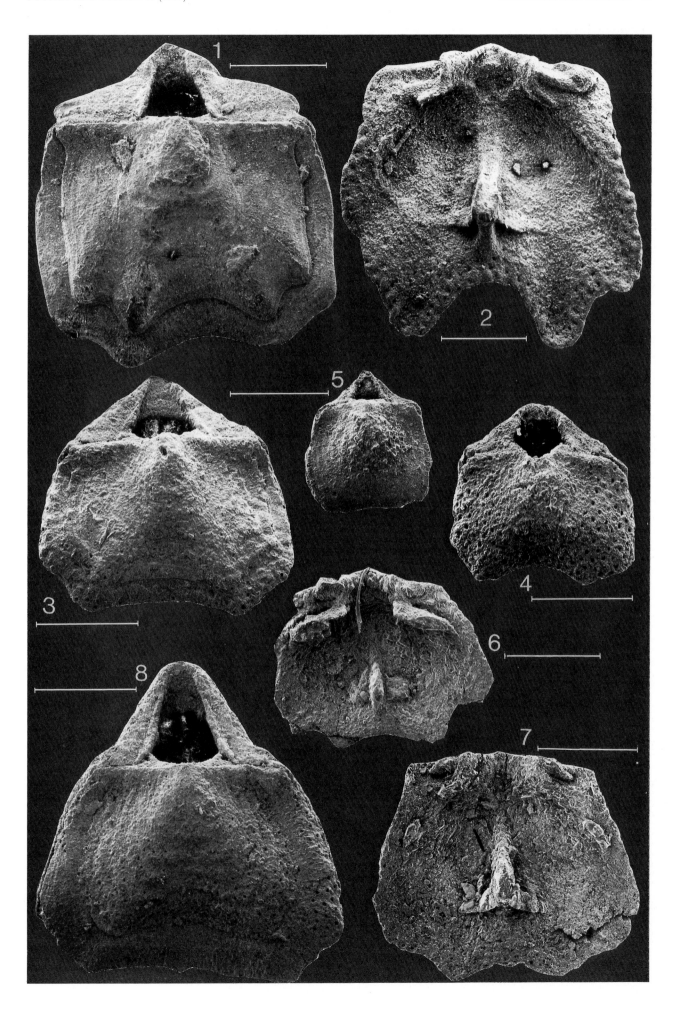

Plate 15

Argyrotheca dorsata (Nielsen 1928) and *Argyrotheca* cf. *faxensis* (Posselt 1894).

☐ 1. *Argyrotheca dorsata*. Complete adult specimen. Sample NK 22, Lower Danian. MGHU 16981. ☐ A. Dorsal view. ☐ B. Lateral view.

☐ 2. *Argyrotheca dorsata*. Fragmented two-valved specimen. Sample NK 24, Lower Danian. MGUH 16982. ☐ A. Ventral view revealing interior of dorsal valve. ☐ B. Lateral view showing dorsal median septum in profile.

☐ 3. *Argyrotheca* cf. *faxensis*. Adult ventral valve. Sample NK 26, Lower Danian. MGUH 16983. ☐ A. Interior. ☐ B. Lateral view showing median septum in profile.

☐ 4. *Argyrotheca* cf. *faxensis*. Slightly damaged two-valved adult specimen in dorsal view. Sample NK 26, Lower Danian. MGUH 16984.

☐ 5. *Argyrotheca* cf. *faxensis*. Interior of adult ventral valve. Sample NK 26, Lower Danian. MGUH 16985.

☐ 6. *Argyrotheca* cf. *faxensis*. Complete juvenile specimen in dorsal view. Sample NK 26, Lower Danian. MGUH 16986.

Scale bars 0.5 mm.

Plate 16

Argyrotheca vonkoeneni (Nielsen 1928).

☐ 1. Interior of medium-sized dorsal valve. Sample NK 24, Lower Danian. MGUH 16987.

☐ 2. Interior of medium-sized ventral valve. Sample NK 24, Lower Danian. MGUH 16988.

☐ 3. Fragmented adult two-valved specimen in ventral view showing cardinalia on dorsal valve. Sample NK 29, Lower Danian. MGUH 16989.

☐ 4. Complete juvenile specimen in dorsal view. Sample NK 24, Lower Danian. MGUH 16990.

☐ 5. Adult dorsal valve. Sample NK 24, Lower Danian. MGUH 16991. ☐ A. Interior. ☐ B. Oblique lateral view showing dorsal median septum in profile.

Scale bars 0.5 mm.

Plate 17

Argyrotheca aff. *faxensis* (Posselt 1894), *Argyrotheca armbrusti* (Schloenbach 1866), and *Argyrotheca* cf. *vonkoeneni* (Nielsen 1928).

□ 1. *Argyrotheca* aff. *faxensis*. Adult dorsal valve. Collected and determined by M. Meyer (1963) as *Megathyris* spp. The specimen is from an unknown locality representing the Guelhem Chalk of Middle Danian Age (Felder 1975), the Limbourg area, the Netherlands. MGUH 16992. □ A. Interior. □ B. Oblique lateral view.

□ 2. *Argyrotheca armbrusti*. Complete adult specimen. Sample NK 17, Lower Danian. MGUH 16993. □ A. Dorsal view. □ B. Lateral view. MGUH 16993.

□ 3. *Argyrotheca armbrusti*. Complete juvenile specimen in dorsal view. Sample NK 17, Lower Danian. MGUH 16994.

□ 4. *Argyrotheca armbrusti*. Interior of juvenile dorsal valve. Sample NK 17, Lower Danian. MGUH 16995.

□ 5. *Argyrotheca* cf. *vonkoeneni*. Adult dorsal valve. Collected and determined as *Megathyris* spp. by M. Meyer (1963). The specimen is from a hardground in the Guelhem Chalk from an unknown locality of the Limbourg area, the Netherlands. The Guelhem Chalk is of Middle Danian Age (Felder 1975). MGUH 16996. □ A. Interior. □ B. Oblique lateral view showing median septum in profile.

Scale bars 0.5 mm.

Plate 18

Aemula inusitata Steinich 1968, *Dalligas nobilis* Steinich 1968, and *Scumulus inopinatus* Steinich 1968.

□ 1. *Aemula inusitata*. Ventral adult valve. Dorsal valve of this specimen represented by 2. Sample NK 1, Upper Maastrichtian. MGUH 16997A. □ A. Exterior, showing characteristic surface sculpture. □ B. Enlargement of surface sculpture shown in A.

□ 2. *Aemula inusitata*. Interior of adult dorsal valve showing recrystallized spicular skeleton. Ventral valve of this specimen represented in 1. Sample NK 1, Upper Maastrichtian. MGUH 16997B.

□ 3. *Aemula inusitata*. Interior of medium-sized ventral valve showing a very low beak and a fragment of the dorsal valve. The dorsal valve of this specimen represented by 4. Sample NK 25, Lower Danian. MGUH 16998A.

□ 4. *Aemula inusitata*. Interior of medium-sized dorsal valve showing an amphithyridid foramen and a very shallow median septum. Ventral valve of this specimen represented by 3. Sample NK 25, Lower Danian. MGUH 16998B.

□ 5. *Aemula inusitata*. Two-valved adult specimen. Sample NK 25, Lower Danian. □ A. Dextral half in ventral view showing surface sculpture, amphithyridid foramen and median septum. MGUH 16999A. □ B. Sinistral half in ventral view. MGUH 16999B.

□ 6. *Dalligas nobilis*. Interior of beak from adult ventral valve. Sample NK 5, Upper Maastrichtian. MGUH 17000.

□ 7. *Dalligas nobilis*. Interior of fragmented adult dorsal valve showing characteristic low median septum. Sample NK 1, Upper Maastrichtian. MGUH 17001.

□ 8. *Scumulus inopinatus*. Complete juvenile specimen in dorsal view. Note the characteristic amphithydirid foramen. Sample NK 3, Upper Maastrichtian. MGUH 17002.

□ 9. *Scumulus inopinatus*. Complete adult speciment in dorsal view. Note the amphithydirid foramen. Sample NK 1, Upper Maastrichtian. MGUH 17003.

Scale bars 0.5 mm except where otherwise stated.

Plate 19

Platidia sp., *Dalligas* sp., and *Scumulus?* sp.

☐ 1. *Platidia* sp. Complete adult specimen in dorsal view. Sample NK 21, Lower Danian. MGUH 17004.

☐ 2. *Platidia* sp. Interior of fragmented adult dorsal valve showing broken crura, and median septum and amphithyridid foramen. Sample NK 22, Lower Danian. MGUH 17005.

☐ 3. *Platidia* sp. Interior of broken adult dorsal valve showing crura and median septum. Sample NK 25, Lower Danian. MGUH 17006.

☐ 4. *Dalligas* sp. Interior of fragmented dorsal valve showing broken crura and short median septum. Two shallow crests run from the median septum in a posterio-lateral direction. Sample NK 22, Lower Danian. MGUH 17007.

☐ 5. *Dalligas* sp. Complete juvenile specimen in dorsal view. Sample NK 25, Lower Danian. MGUH 17561.

☐ 6. *Scumulus?* sp. Complete adult specimen. Sample NK 5, Upper Maastrichtian. MGUH 17562. ☐ A. Dorsal view. ☐ B. Lateral view.

☐ 7. *Scumulus?* sp. Complete juvenile specimen in dorsal view. Sample NK 6, Upper Maastrichtian. MGUH 17563.

Scale bars 0.5 mm except where otherwise stated.

Plate 20

Leptothyrellopsis sp. and *Kingena pentangulata* (Woodward 1833).

☐ 1. *Leptothyrellopsis* sp. Fragmented two-valved specimen in dorsal view. Sample NK 8, Maastrichtian–Danian boundary clay. MGUH 17564.

☐ 2. *Leptothyrellopsis* sp. Complete juvenile specimen in dorsal view. Sample NK 4, Upper Maastrichtian. MGUH 17565.

☐ 3. *Leptothyrellopsis* sp. Interior of adult dorsal valve showing recrystallized spirolophe(?) and median septum. Sample NK 4, Upper Maastrichtian. MGUH 17566.

☐ 4. *Leptothyrellopsis* sp. Juvenile dorsal valve. Sample NK 8, Maastrichtian–Danian boundary clay. MGUH 17567. ☐ A. Interior showing median septum and broken brachidium. ☐ B. Oblique lateral view showing median septum in profile.

☐ 5. *Kingena pentangulata*. Complete juvenile specimen in dorsal view. Sample NK 5, Upper Maastrichtian. MGUH 17568.

☐ 6. *Kingena pentangulata*. Interior of broken juvenile dorsal valve showing brachidium and recrystallized lophophore in early precampagiform stage. Sample NK 5, Upper Maastrichtian. MGUH 17569.

Scale bars 0.5 mm except where otherwise stated.